THE
HARMONIC
ORIGINS
OF THE
WORLD

"In this book the author reveals himself as the natural successor to Ernest McClain. As McClain before him, Heath has realized the extent to which the natural harmony of music binds into one the 'within' and 'without' of man's world. He introduces us first to the observable recurrence of numbers underlying ancient astronomical sightings and then skilfully reveals the connection with the harmonic numbers of the sexagesimal system discovered by the Sumerians and Babylonians. Underpinning ideas with superb graphics and skilful numerical tables, he shows the ancient scribes, priests and 'gentlemen of leisure' in the Aristotelian sense, to be most subtle—in many cases far more so than we who work so hard to understand them. The progression by which he reveals his thesis is impressive."

PETE DELLO, SINGER-SONGWRITER,
COMPOSER, AND MUSICOLOGIST

THE
HARMONIC
ORIGINS
OF THE
WORLD

SACRED NUMBER AT THE
SOURCE OF CREATION

RICHARD HEATH

Inner Traditions
Rochester, Vermont • Toronto, Canada

Inner Traditions
One Park Street
Rochester, Vermont 05767
www.InnerTraditions.com

Library of Congress Cataloging-in-Publication Data

Names: Heath, Richard, 1952– author.
Title: The harmonic origins of the world : sacred number at the source of creation / Richard Heath.
Description: Rochester, VT : Inner Traditions, 2018. | Includes bibliographical references and index.
Identifiers: LCCN 2017028598 (print) | LCCN 2017057856 (e-book) ISBN 9781620556122 (pbk.) | ISBN 9781620556139 (e-book) |
Subjects: LCSH: Cosmology. | Number theory—Miscellanea. | Creation. | Harmony of the spheres. | Religion, Prehistoric.
Classification: LCC BD511 .H428 2018 (print) | LCC BD511 (e-book) | DDC 113—dc23
LC record available at https://lccn.loc.gov/2017028598

Printed and bound in the United States by P. A. Hutchinson Company

10 9 8 7 6 5 4 3 2 1

Text design and layout by Debbie Glogover
This book was typeset in Garamond Premier Pro with Posterama Text and Gill Sans used as display typefaces
Images 7.1 and 7.3 appear courtesy of their creators through https://creativecommons .org/licenses/by/3.0/us/legalcode

To send correspondence to the author of this book, mail a first-class letter to the author c/o Inner Traditions • Bear & Company, One Park Street, Rochester, VT 05767, and we will forward the communication, or contact the author directly at **sacrednumber@gmail.com**, or through **www.richardheath.info**.

To
Ernest G. McClain
Cryptographer to the Gods
(1918–2014)

Contents

∾❦◎❧∿

Preface

My own findings regarding planetary harmony lacked a proper context until I came to understand the work of Ernest G. McClain and collaborate with him under the auspices of Duane Christensen's Bibal group, many of whose members were Ernest's highly various friends and long-term collaborators. Ernest sought to write his version of this book, then called *Brave New World*. Instead, there was a period of acclimatization and analysis as to the deeper implications within his diagrams and his methods of working. Out of the blue I wrote him a program through which anyone can survey his views of the harmonic domain, while that domain and his friendship started to change how I perceived it.

Previous books of mine have also drawn on bodies of work not generally accepted by academics and hence were published to a more general audience, those interested in Earth mysteries and sacred numbers. A prominent influence was a leading author, John Michell, with whom I corresponded largely about ancient metrology. Michell and his friend John Neal provided me with the content for a self-propelled education in the now suppressed science of measures, which has been crucial here in analyzing sites such as the Parthenon, Marduk's ziggurat, and the Mexican city of Teotihuacan, as to harmonic codes built using ancient measures.

I wrote my first book (*Matrix of Creation*) after a decade of calculation. This work was initiated by my brother Robin, and it focused on how the megalithic probably studied astronomy through their monuments and geometrical methods, then stumbling on the harmonic ratios crucial to this book. My purpose became the work of writing and researching a number of mystery areas with an eye to the story of their human development, a story not sought by historians who ignore fringe subjects. My writing resonated with the interests of long-term friend and occasional mentor Anthony G. E. Blake, on how media journeys enable an implicit order toemerge.

Richard Dumbrill's ICONEA musicology conferences for 2013 and 2014 led to the work found here on harmonic buildings, the Olmec, and orality, while Dumbrill and Carr's new *Music and Deep Memory: In Memoriam to McClain* will have an extensive paper on harmonic allusion in the earliest chapters of the Bible, from which chapter 3 developed. The Bible material is a concrete example of what was possible for this book, and since then other topics, undocumented by McClain, have been completed including the location of musical scales on harmonic "mountains" with help from my Bibal friend Pete Dello.

Having published five books with Inner Traditions I recognize, from a media perspective, how contact with technical specialists and their general ethos has influenced my published work for the better. The copy editor for this book, Cannon Labrie, and project editor, Jennie Marx, cannot have found it easy handling new semitechnical material, yet their work triggered a significant—and, I think, beneficial for the reader—restructuring of the book.

Researching illustrations has become easier online with Creative Commons and Wikipedia, while rights to use have become harder if not impossible to progress in some cases. Michael D. Coe very kindly agreed to my use of his content, as did Carly Rustebakke on behalf of the Linda Schele collection and David Schele.

Before he died, Ernest McClain gave his permission to use any of his work, helping me to create sections of chapter 4 on Greece. But I could not rest on McClain's laurels and, where possible, I have instead expanded on his own body of examples for harmonic coding employed in the ancient world. One should not assume McClain would have approved of my work but, in the service of my own findings regarding planetary harmony, his own case for harmonic allusion in ancient literature is further validated—for a widespread religious harmonism having existed before the current era.

The Significance of Planetary Harmony

Over the last seven thousand years, hunter-gathering humans have been transformed into the "modern" norms of city dwellers through a series of metamorphoses during which the intellect developed ever-larger descriptions of the world. Past civilizations and even some tribal groups have left wonders in their wake, a result of uncanny skills—mental and physical—that, being hard to repeat today, cannot be considered primitive. Buildings such as Stonehenge and the Great Pyramid of Giza are felt to be anomalous because of the mathematics implied by their construction. Our notational mathematics only arose much later, and so, a different mathematics must have preceded ours.

We have also inherited texts from ancient times. Spoken language evolved before there was any writing with which to create texts. Writing developed in three main ways: (1) pictographic writing evolved into hieroglyphs, like those of Egyptian texts, carved on stone or inked onto papyrus; (2) the Sumerians used cross-hatched lines on clay tablets to make symbols representing the syllables within speech; cuneiform allowed the many languages of the ancient Near East to be recorded, since all spoken language is made of syllables; (3) the Phoenicians developed the alphabet, which was perfected in Iron Age Greece through identifying more phonemes, including the vowels. The Greek language enabled individual writers to think new thoughts through writing down their ideas, which was a new habit that competed with information passed down through the oral tradition. Ironically though, writing down oral stories allowed their survival, as the oral tradition became more or less extinct. And surviving oral texts give otherwise missing insights into the intellectual life behind prehistoric monuments.

The texts and iconography of the ancient world once encrypted the special numbers used to create ancient and prehistoric monuments, using a numeracy that modeled the earth and sky using the invariant numbers found in celestial time, and in the world of number itself. Oral stories spoke from a unified construct, connecting the people to their gods. Buildings were echoes of an original Building, whose dimensions came to form a canon within metrology, the ancient science of measure. But the language of the gods within this Building was seen to be that of musical tuning theory, the number science that concerns us here. The gods in question are primarily the planetary and calendric periods seen from Earth, and it was only through the astronomy associated with the earliest megalithic buildings that the ancient mathematics could have naturally evolved.

To see musical harmony in the sky, time was counted as lengths of time between visible astronomical events such as sunrise, moon set, or full moon. Geometry evolved to set alignments to horizon events, such as the solstice sun, or to place long lengths of day counting within geometries such as the trigonometric triangle. Megalithic astronomy (chapter 1) consisted of a set of quantified lengths of time and the geometrical relationship between them. It would have discovered that some of the ratios between time periods were especially simple: most significantly, the two outer planets, Jupiter and Saturn, were related as 9/8 (a musical whole tone interval) and 16/15 (a semitone interval) to the lunar year. In each ratio the lunar year is the denominator and the planetary synods are the numerators. If we make the denominators the same (by multiplying the ratios by each other's denominator) we obtain (times 15) 135/120, and (times 8) 128/120. Because the lunar year is 12 lunar months long, the lunar month must comprise ten subunits of time; the Jupiter synod must be 13.5 months long, and Saturn's synod must be 12.8 months long.

The idea that astronomy could have caused the ancient world to have any great interest in musical tuning theory runs against the standard musicological model of history in which it was the making of music that drove the Babylonian tuning texts to appear on cuneiform tablets from Nippur and other places. However, lists of regular numbers and tables of reciprocals counting down from sixty to the power of four (12,960,000) hardly seem relevant to practical music. The number 12,960,000 is a significant number in my work, belonging to Venus, the bright planet of the inner solar system, in its synod relative to the lunar year. The number is a large one because she is higher in "heaven," becoming Quetzalcoatl in the Olmec's cult of astronomical time inherited by the Mayan and Aztec cultures (chapter 8). Tuning theory must have found its way to Mexico before the devastating Bronze Age collapse circa 1200 BCE, a date when Mexico's likely contact with the eastern Mediterranean would have ceased. The future of European tuning theory in the ensuing Iron Age then lay in the hands of the Archaic Greeks (Homer and Hesiod) and, surprisingly, the Jewish school responsible for the early Bible (chapter 4).

Whenever civilizations fall they pass on information. When megalithic astronomy died, it bequeathed the idea that the planets were gods related to the moon through musical harmony, also leaving the ancient world its metrology. When temples were built or stories to the planetary gods passed on, these could express musical numbers as ratios and lengths within architecture, iconography, and myth. In classical Greece, the power of writing had won over the oral world whereupon Athens enshrined musical harmony in the Parthenon and in Plato's writings about the ancient tradition of musical tuning theory (chapter 4).

I first noticed the musical resonances (of Jupiter and Saturn to the lunar year) in 2000, for which I could find no traditional setting except mythology.[1] The extensive works of the Pythagorean tradition for instance, concerned with planetary harmony, are complex and appear more influenced by Greek mathematics than by the ancient world. After some decades though, understanding came through the work of Ernest G. McClain and through my collaboration with him in the last years of his life. These outer planetary resonances slotted perfectly into

McClain's frameworks for ancient tuning theory. The primary sources for McClain's work were the surviving texts of the ancient world but his key to these texts was Plato's dialogues, for which he had provided a definitive interpretation, as being a cryptic textbook for ancient tuning theory.[2] Ernest McClain found harmonic numbers (which only have factors of 2, 3, and 5) referred to (as if arbitrarily) in various guises within ancient stories, allowing the initiated to reconstitute a much larger array of harmonic numbers belonging to the god or to a spiritual locale that the story was intended to animate.* In his popular work *The Myth of Invariance,* McClain recreated many otherwise hidden harmonic worlds from number references within texts from India, Mesopotamia, Egypt, Greece, and the New World.

It became obvious to me that the common denominators (see ratios of the Jupiter and Saturn synodic periods mentioned above) would "place" them in the corner of McClain's "holy mountains."† More and more "characters" from astronomical time started appearing "on the mountain," in parallel with McClain's own interpretations from the Bible, Homer, Babylonian texts, the Rg Veda, and so on (for example, see figure 8.7, page 177).

The astronomical significance of harmonic numbers left in ancient texts explains the mystery of why they should be there in the first place, and it confirms the important role texts have played in carrying a whole system of knowledge from an oral tradition. In their heyday, "texts" were only in the heads of reciters and listeners of all sorts—some hearing a good story and others learning new facets of harmonic knowledge. Such a tradition evidently thought the world had come into existence due to musical harmony (chapter 6) and that relationships to the gods were organized according to harmonic laws. Indeed, the astrology that so obsessed the Babylonians was probably rooted in a harmonic

---

*This resembles the aboriginal habit of recitations before painted caves or cliffs, where an initiate recounts the story illustrated by the paintings, thus "joining the dots." Our word *esoteric* perhaps derives from this practice of leaving cryptic clues within texts, in this case linking to musical tuning theory.

†His arrays of regular numbers that could be inferred from a single number limiting the array and that serve as the high *do* note in an octave.

model of fate involving planets and calendars. One can see how stories such as Gilgamesh reveal that the Sumerians knew of it (ca. 3000 BCE), placing planetary gods like Inanna/Ishtar/Venus in heroic stories that make better sense if referring to "holy mountains" (chapters 2 and 7) located in a harmonic heaven.

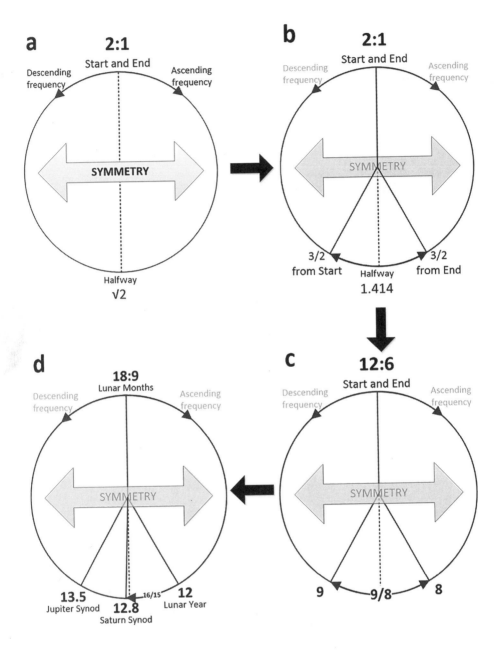

Figure PI.I. How the octave domain (*a*), which is symmetrical in the logarithmic domain we hear, becomes (*b*) Plato's Demiurge consisting of the descending and ascending fifths (3/2), then (*c*) multiplies the octave by 6 to rationalize these to pitches values 8 and 9 within octave 6:12, and (*d*) this then fits the relative frequency (in months) of the lunar year, Jupiter and Saturn synods.

PART 1

RECOVERING LOST KNOWLEDGE OF THE WORLD SOUL

Plato proposed a World Soul that took the form of a musical octave—having been given that form by the Demiurge, a creator god—bringing harmony to the Earth's surroundings. The octave is so named because in modal music there are only eight tones, and the beginning and ending tones differ in frequency by a factor of exactly 2. In the act of creation there were two further tones within the octave— mirror images of each other. This mirroring, called *symmetry,* is crucial within all created octaves as tones arise that are reciprocal to each other and to both ends of the octave, like twins (see figure P1.1).

To see the octave's true form, one must visually mimic the ear's ability to see all identical intervals as equal, and to use the eye to see all octaves as circles going clockwise from top (= 1) until the top is reached again. At the bottom of the circle lies a possible tone that is equidistant from the top, traveling either clockwise or counterclockwise. Since each path is equal, the value of this tone must be the square root of 2, whose square would then be the multiplication by 2 involved in doubling (see *a* in figure P1.1).

The Demiurge must decide on *what* is going to be doubled in size to form the octave, in the context of *what* is going to populate this tone circle. Chapter 1 reveals the lunar month was used by the Demiurge for the purpose of calibrating his octave as follows.

Plato's Demiurge took the lower value of the octave (1), divided it by two and added the half to the lower value obtaining the tone 1.5 or 3/2, which in music is a fifth. He also divided the doubled value of the octave (2) by three and subtracted that to obtain the tone 1.$\underline{333}$ = 4/3 from the start (see *b* in figure P1.1). To achieve this with all integer values, the octave would have to be scaled up by a factor of 6 to having 6 to 12 linear units. The four tones of the octave are then 6:8::9:12 units, with 6 and 12 at the top, as is shown in *c* in figure P1.1. However, instead of using the *simplest possible* octave of 6:8::9:12, the real Demiurge rescaled all the numbers by 3/2, making them 9:12::13.5:18 in lunar months (or in

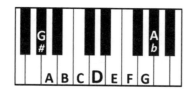

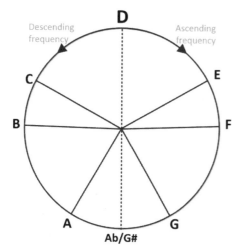

Figure P1.2. The modern note names and keyboard reflect the significance of symmetry to the octave domain.

integer half months, 18:24::27:36), noting that twelve months make up the lunar year (see *d* in figure P1.1).

The tones of the octave need to be introduced by their modern note names. Because the tone circle is inherently symmetrical, the tone names used in this book are aligned with those note names used for our modern keyboard's twelve different tones (the scheme adopted by Ernest McClain). The top and bottom notes of the octave should be D, since the piano key for D is the *only* symmetrical white note, surrounded by left-right symmetry of black and white keys. The only symmetrical *black* note is Ab/G#, and this is numerically the square root of 2 referred to above and shown in figure P1.1*a*.

Using note names, the octave is then between D as 9 lunar months and D as 18 lunar months. The two created tones within the octave are G as 12 lunar months (the lunar year) and A as the Jupiter synod of 13.5 lunar months. While the Jupiter synod, seen only from the Earth, was probably not changeable by the Demiurge, the other three tones of D, G, and octave D are all consequential to a lunar orbit that was increasing its distance from the Earth due to tidal interaction. The increasing lunar orbit might have allowed this harmonious octave relationship to develop as follows.

A clue is found in Saturn, the next planet out from Jupiter, whose synod is now 12.8 lunar months so that these two outer planetary synods are inherently related, irrespective of the moon, by the interval 135/128. The lunar year is related to Saturn as 128/120, which reduces to 16/15, an exact semitone from the lunar year and hence on the lower point of the octave's line of symmetry, opposite D (see *d* in figure P1.1). Saturn is traditionally termed a demiurge, and this lower point is considered the location of the creative force or fire god, a point of sacrifice and regeneration.

Plato's World Soul coalesced into this harmonic pattern about 200,000 years ago, a time frame parallel to the evolution of modern man (see figure P1.3). It therefore appears Plato's Demiurge referenced a lost body of knowledge that understood that the major planets were in harmonic relationship to our moon.

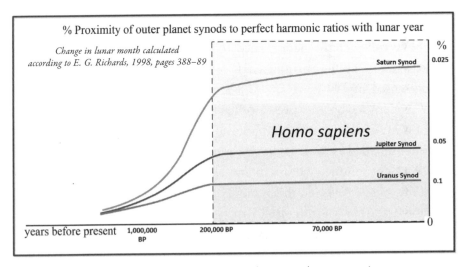

Figure P1.3. Countdown to *Homo sapiens* as the outer planets come into resonance with the lunar year. Time in years runs left to right B.P., and the vertical axis is the percentage accuracy of the harmonic intervals of the three nearest outer planets, as this set in with the arrival of *Homo sapiens*, between 0.1 percent (Uranus) and 0.025 percent (Saturn) with respect to the lunar year.

1

Climbing the
Harmonic Mountain

Humans of the Neolithic period* built a large number of megalithic monuments. A representative number of these have survived the eroding effects of natural and human forces, which has allowed the design of these megalithic monuments to be assessed by some, including myself, as demonstrating a very sophisticated and advanced understanding of astronomical time periods. But such interpretations are effectively ignored by academic archaeology since they imply the existence of exact sciences and numeracy in the Neolithic, which is only thought of as innovating agrarian lifestyles. There is some consensus that monuments had some astronomical uses: for example, tracking the seasonal motion of the sunrise (or sunset) to form a solar calendar, or tracking lunar phases using daily horizon events. But evidence supporting an exact and unique form of *megalithic* astronomy is resisted because counting time periods and constructing numerate monuments required some form of mathematics.

---

*New Stone Age, circa 10,000 BCE to 2000 BCE.

THE CYCLICAL NATURE OF
THE CELESTIAL WORLD

The natural unit of time on Earth is the day, and the counting of days would lead one to see that there are 59 days between three full moons. But this generates a problem when we ask whether the number 59 could have had any meaning in a pre-arithmetic society, which the Neolithic appears to have been. Marks on bones from the Stone Age suggest that counting occurred but without thought for number as an abstract or an externalized entity. Instead, the number of day marks probably led to the notion of numbers as *equal to a length* containing "that many," as a number.

The sun offers a more direct form of calendar than the moon, corresponding to the new needs of seasonal farmers in the Neolithic and hence the agricultural context within which megalithic monuments were being built, in some regions. The sun appears to move north (in the Northern Hemisphere) toward summer and south toward winter. Looking south, the noonday sun gets higher in the summer and lower in the winter and, as a rule of thumb, this is what makes summer hotter than winter. Summer moves the location of sunrise and sunset toward the northeast and northwest, respectively, and by a decreasing amount per day. At the spring and autumn equinox, sunrise and sunset are exactly due east and west in most latitudes, and in winter they move south so as to be (equally) southeast and southwest. Their furthest positions are the solstices of summer and winter, after which they head back to their equinoctial locations, east and west.

This movement of the sun to north and south therefore causes sunrises and sunsets from the center of any megalithic site to move along the horizon between two extreme positions during the year—the summer and winter solstice—and thus, the horizon to east or west of a megalithic site formed a natural solar calendar. At megalithic sites one often finds that some key points on the horizon coincide with sunrise or sunset for a solstice or the equinox, which means that the site was likely selected to have that characteristic—namely, that the view from a fixed observation point at the site naturally aligned to the sun on the horizon at a moment significant to a megalithic solar calendar. This alignment to the natural landscape could then be extended by providing man-made features, typically standing stones, which could then be viewed from the megalithic site's observation point to indicate where the sun would stand at another point of the solar year. Between the solstice "extremes," each such marker to the horizon in fact marks two days in the year since the sun travels both north and south during the year and will stand above that marker twice in the year, during the waxing and waning of the sun's influence.

The moon's behavior on the horizon is a more complex phenomenon, so it is easier to track over a smaller time frame using its waxing and waning phases, these defining the synodic lunar month and the moon's changing illumination by the sun. The full moon and the different moon halves are particularly distinct phases, the growing half waxing and the diminishing half waning.

Thus, the fundamental observing framework for the sun was the horizon, while for the moon, the phases of the lunar month were easier to observe. Such observations provide no real explanation for the building of complex megalithic monuments. If only such basic observational astronomy was involved, what drove the complex design of monuments? Two types of explanations arose in the twentieth century, once the sites started to be accurately surveyed and dated.

The first type of explanation suggests that the building of monuments was driven by an already existing cultural framework that some would call religious, or at least involving death and the dead. It gave significance to dates within the solar year, or parts of the lunar month, within which commemorative rites were practiced or celebrated at the monuments. Such proposed explanations therefore saw megalithic monuments as symptomatic of preexisting religious frameworks tied to celestial events.

The second type of explanation was that the monument builders were an intellectual elite of their day who achieved significant feats of astronomical understanding, self-evident in the forms that the monuments took. Prominent was a proposed explanation, based on factual evidence, that the monuments embodied a metrological and geometric competence enabling a pre-arithmetic Neolithic Age to solve arithmetical problems and resolve astronomical facts not thought knowable to Stone Age people, but easily graspable using our own scientific methods.

The first explanation, that monuments were venues involving proto-religious calendric events, surprisingly accepted by today's specialists, lacks concrete evidence that would prove such a usage as having been primary for megalithic monuments. In other words: to accept time-factored ritual as the raison d'être for megalith-ism in general is a leap of faith that can neither be proven nor refuted. This interpretation therefore forms a cul-de-sac for rational thought and prevents the search for why complex megalithic monuments were built.

Contrast that with the second explanation, that the monuments involved a sophisticated combination of astronomy, geometry, and metrology, which has been criticized as being far too precise and the result of overenthusiasm by individuals predisposed to find their own meanings in these monuments, meanings they are therefore "selecting for," thus rendering them not objective. However, unlike the ritual explanations for which there is no concrete evidence, the metrological explanation can be tested through measuring the sites and so may be refuted, in the Popperian sense, as to its validity. The problem lies in cultural resistance to revising the standard model of history. Those proposing a ritual explanation are deselecting explanations that the megalithic was an advanced and highly numerate culture because modern numeracy first developed in the ancient Near East, before the development of our geometrical methods for solving astronomical problems.

This unfortunate rejection is fortunately irrelevant to our concerns here. What follows is an abbreviated account of how the megalithic builders, having conquered the challenge of gaining accurate knowledge of celestial time periods, were thus led to the discovery of musical harmony between planetary periods. Musical harmony is based on the three earliest prime numbers: 2, 3, and 5. The discovery of celestial harmonies may have ended the megalithic project and initiated the new religious, literate, and mathematical civilizations in the Near and Far East that, through their written and other records, gave us our earliest histories.

THE MOON AS A
MEASURE OF ALL THINGS

The geometrical comparison of lengths of time can only have arisen through the adoption of a standard length for counting, where one day was counted a standard unit of length such as, for example, the inch. Robin Heath and I found, within Carnac's Le Manio quadrilateral, a monument in which a special trigonometric triangle had been defined.[1] Its base is four units and its shortest side one unit: that is, it is a triangle based on a 4-by-1 rectangle, where the triangle's longest side is the rectangle's diagonal. The two longer sides of this triangle were "day-inch" counts for three lunar years (the base) and for three solar years (the hypotenuse), so that the unit sides of the squares in the foursquare rectangle (and the short side of the triangle) were nine lunar months long, in day-inches. If instead one built a foursquare rectangle with a unit side three rather than nine lunar months long, the base and diagonal would be the day-inch counts for a single lunar and a single solar year. Once this geometry had been found, it could be reproduced like a calendar.

In modern times this triangle, base 12 lunar months (the lunar year) and short side 3 lunar months, was first established as having been a useful geometry by my brother around 1990.[2] By noting that the Station Stone rectangle at Stonehenge was 12 units by 5 units, a division of the 5 units into 3 units and 2 units allowed an intermediate

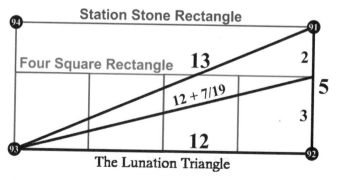

Figure 1.1. Robin Heath's lunation triangle in the context of the 12-by-5 Station Stone rectangle at Stonehenge and that triangle's "foursquare" rectangle evident at Le Manio quadrilateral near Carnac, Brittany.

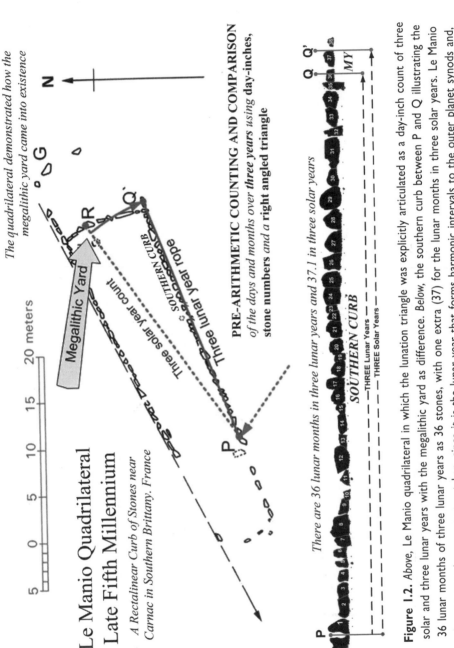

The quadrilateral demonstrated how the megalithic yard came into existence

N

20 meters

Le Manio Quadrilateral
Late Fifth Millennium

A Rectalinear Curb of Stones near Carnac in Southern Brittany. France

Megalithic Yard

SOUTHERN CURB

Three lunar year rope

Three solar year count

PRE-ARITHMETIC COUNTING AND COMPARISON
of the days and months over three years using day-inches,
stone numbers and a right angled triangle

There are 36 lunar months in three lunar years and 37.1 in three solar years

SOUTHERN CURB

THREE LUNAR YEARS
THREE SOLAR YEARS

MY

Figure 1.2. Above, Le Manio quadrilateral in which the lunation triangle was explicitly articulated as a day-inch count of three solar and three lunar years with the megalithic yard as difference. Below, the southern curb between P and Q illustrating the 36 lunar months of three lunar years as 36 stones, with one extra (37) for the lunar months in three solar years. Le Manio shows a very important step taken, since it is the lunar year that forms harmonic intervals to the outer planet synods and, right-angled triangles were the available means by which planetary synods could be compared with the lunar year.

hypotenuse to be formed that, compared to the base of 12 units, would then be 12.368 units long. This hypotenuse generated the number of lunar months within a solar year (12.368 months), and so the triangle geometrically represents the relationship between the lunar year and solar year, in lunar months. This triangle was therefore called the lunation triangle, since its formation appears to represent how the megalithic astronomers had resolved the great challenge of forming a simple sun-moon calendar, successfully integrating lunar months within the solar year as a quantifiable though endlessly slipping ratio, as one might find between two gears of a car.

At Stonehenge the lunation geometry can only be inferred as a potential meaning for the rectangular dimensions 12 by 5. The metrology of the Station Stone rectangle's longest side (figure 1.1) is 8 × 12 megalithic yards, numerically correct for the 12 side of the lunation triangle. If used as a calendar (where one "counts" outside of the present moment) the monument could (*a*) be counting in months rather than days over (*b*) the eight-year periodicity for a first good solar return on the horizon (where the sun rises in exactly the same place on the horizon, on the same day of the year it had in a previous year), and its units are 8 megalithic yards.

Le Manio preceded the Station Stone rectangle by at least five hundred years, so its quadrilateral could be closer to the moment at which astronomers discovered the foursquare geometry of the sun and moon (this construction being the best and easiest way to reproduce the lunation triangle). Another key discovery at Le Manio appeared as the difference in counted length between the solar and lunar years, over three years; in day-inches it is the megalithic yard of 32.625 inches, a result that probably reveals the origin of the megalithic yard as a measure then used throughout the megalithic period to count lengths, within monuments such as the Station Stone rectangle, in months. It is also true that, by using a slightly smaller megalithic yard* of 19/7 feet to count lunar months, the fractional part of the lunation triangle's hypotenuse (the lunar months in a solar

---

*The astronomical megalithic yard.

year) of 12.368 megalithic yards could be made a rational fraction since 0.368 = 7/19 (lunar month).* Further details about the Manio quadrilateral and the lunation triangle can be found in Richard and Robin Heath, *The Origins of Megalithic Astronomy as Found at Le Manio.*

A key feature of the lunation triangle calibrated in megalithic yards is that, over a single year, another key unit of length emerges: the English foot of twelve inches, as the difference between the solar and lunar years. It also generates the royal cubit, of 12/7 feet, as the difference between the lunar and eclipse years, which added to one foot equals 19/7, the astronomical megalithic yard. It appears that, in the move to counting lunar months in megalithic yards rather than day-inches, the metrology of the late megalithic and of later ancient buildings became founded on the English foot and fractional variations of it, such as 12/7 feet;[3] that is, all ancient measures are related to the foot through one or more rational fractional conversions. If the megalithic yard was the measure first derived from the difference between the solar and lunar years, then it follows that only the megalithic astronomers could have given birth to such a science of measure and geometry that reduced the complexities of horizon astronomy to the kind of predictive certainty seen in the Antikythera mechanism (using gears by the time of classical Greece) or the Mayan calendar (using a long count by the time of the Olmec in Mesoamerica). Such anachronisms should be pondered further as being due to megalithic science.

HARMONIC TRANSFORMATION FOR MEGALITHIC ASTRONOMY

It turns out that counting full moons using megalithic yards was a remarkably fortunate step, providing both a simpler way of counting

---

*Clearing of fractions involves multiplication to remove a ratio's denominator. It becomes especially important here when harmonic limits must be multiplied to include otherwise fractional time periods.

time periods and exposing hidden celestial relationships.* Counts could be reduced in numerical size over their equivalent day-inch counts and, more importantly, the lunar year could be compared to the periodicities of Jupiter and Saturn to reveal harmonious interval ratios. Megalithic astronomers gained an early experience of musical ratios that allowed them to notice how prime-number factors of 2, 3, and 5 were *always* present in musical ratios (see below). According to our physics, the giant planets have had a significant gravitational effect on the moon and brought the lunar year into synodic resonance with them. But instead of the invisible action of gravity, the megalithic identified musical intervals as the primary *form* the relationship between the moon and the outer planets took.

However, measuring the synodic periods of Jupiter and Saturn required a new type of observation in order to know where these two "wanderers" were among the stars at night. Every solar year the Earth overtakes these outer planets, which have longer orbits, at which time the sun is opposite them. This causes each outer planet to come to a standstill and then appear to go backward (or "retrograde") for a period until, reaching another standstill, the planet appears to travel forward again until returning to its original position and resuming its usual course. The result is called a loop because there is also some up-and-down parallax, viewed from Earth. The megalithic astronomers would have observed the moon rising above and below the sun's path during each lunar orbit around the Earth. The outer planets similarly loop around the ecliptic every year, and counting months between two loops of the same planet gave the synodic periods of the outer planets and showed Jupiter's synod (398.88 days) was thirteen months plus half a month, (13.5) while Saturn's synod (378.09 days) was twelve months plus four-fifths of a month (12.8) *almost exactly in both cases.*

Using megalithic yards, Jupiter's synod could be rendered as 135 whole units of 1/10 of a megalithic yard, while Saturn's period

---

*The megalithic yard (32.585 inches) is 32/29 of the lunar month (29.53059) in day-inches, a fact that blocked the fact that the original counting of time as length, prior to the use of the megalithic yard, employed day-inch counting. Note that 32/29 is a rational fraction.

could be presented as 128 whole units of the same length. The same unit presents the lunar year as 120 units long. Using tenths of a megalithic yard naturally removed the common denominator, clearing the fractions found in the synods of Jupiter and Saturn when measured in lunar months. (Jupiter needed a half to be cleared, and Saturn needed a fifth, and the product of 2 × 5 being 10 was provided by using units 1/10 smaller than a lunar month. This division of the month was already available within the known subdivision of the megalithic yard into 40 megalithic inches, 4 such inches being 1/10 of a megalithic yard.)

Once the synods of the moon, Jupiter, and Saturn were made integers, in lunar months (as 120, 135, and 128, respectively) the ratios between these three time periods could be found in a simpler way than had the interval between the sun and the moon.* Both of the visible outer planets were revealed as relating to the moon through the—what we would call—harmonic ratios: the 9/8 "tone" interval for Jupiter and 16/15 "semitone" interval for Saturn.

PRIMES AS DIMENSIONS
WITHIN HARMONY

We have now carefully traversed the unlikely but possible route from megalithic astronomy to the quantification of planetary synods, their significant comparison with the lunar year, and the finding of musical ratios between those synods. One further step was required that would establish a prehistoric tuning theory, and this involves the factors found within the numbers obtained. It was clear that 135/120 in its simplest form was 9/8 when all of the shared factors making up 135 and 120 cancel out. For example, Jupiter's 135 equals 9 × 15 while the moon's 120 equals 8 × 15, so that 15 cancels out, leaving 9/8. In contrast, Saturn's 128 equals 16 × 8, so that 8 cancels out leaving 16/15. The lunar year's number (120) contains only the two common factors to be removed from the other two, namely, 15 × 8 = 120.

---

*Namely, 12,000 to 12,368 thousandths of a lunar month, in integers.

Curiosity, as to what primes *were*, would naturally have occurred in the megalithic owing to their need to play with factors within metrological lengths. The simple exercise of counting synodic periods had resulted in the discovery of ratios born purely from the earliest prime numbers, and this would naturally lead to a study of the harmonic primes. Since each prime could not divide into the others, harmonic intervals were moving within a dimensional space as three dimensions that could not mix, made up of the powers of 2, 3 and 5.* Saturn was made up entirely of 2s, the lunar year had a 3 and a 5, while Jupiter had a 27 (= 3 × 3 × 3, or 3^3) and a 5. The prime 5 links the Jupiter synod and lunar year while they differ by 9/8. Following the lead of Ernest McClain (1976), we can therefore allocate two dimensions to show the "state space" within which these three synods sit. Up will be the powers of 5 and across will be the powers of 3. The lunar year and Jupiter synod are "above" Saturn, which has no 5. The lunar year will be diagonal to Saturn, differing by one power of both 3 and 5. Jupiter will be two positions further to the right of the lunar year because of its two further powers of 3, as shown in table 1.1.

TABLE 1.1

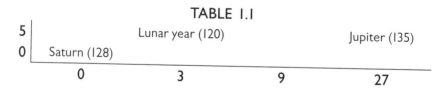

5	Lunar year (120)		Jupiter (135)
0	Saturn (128)		
0	3	9	27

That the ancient world came to musical harmony through astronomical ratios goes against a well-entrenched dogma that systematic experiments in musical harmony—the varying string lengths and noting the more harmonious resulting intervals between tones—came from experiments conducted by the Greek Pythagoras (ca. 600 BCE). There is a natural duality inherent within musical harmony concerning the role of the ear, in hearing harmony, and that of the intellect, in

---

*Prime-number dimensionality was, by the very end of the ancient world, Plato's understanding, and, without separating the primes, tuning theory could not have been constructed by the ancient world (see McClain, *The Pythagorean Plato*).

knowing the numerical rules that govern harmony. Pythagoras is portrayed experimenting with the lengths of vibrating strings, resulting in harmonious musical intervals between strings. While the ear can know harmony directly in the experiments, the numbers found in the corresponding string lengths are the "intellectual" ratios such as 9/8 and 16/15.

I hope the reader can see the strong connections between Pythagoras's experiment in comparing string lengths and the megalithic comparing of lengths of time to show celestial musical intervals. My proposal is that, like most myths, the myth of Pythagoras is not exactly what it seems. The story may well celebrate the original discovery of celestial harmony after which experiments with similar ratios of vibrating strings caused the invention of tuning theory. If so, then the myth of Pythagoras points three thousand years earlier. In any case, acoustic experiments appear to have been conducted by the third millennium BCE in which scribes gave the *names* of harp strings as being the smallest *number* associated with a given scale's set of tones. Pythagoras could not have innovated numerical tuning since Old Babylonians clearly had contact with a numerical tuning theory. However, it is Pythagoras who kept alive the astronomical idea of a harmony of the spheres, and this led to the many later theoretical harmonists, who were the cosmologists of their day (see "The Harmony of the Spheres" in chapter 4), often called Pythagoreans.

A NEW ANATOMY FOR HARMONY

If Pythagoras understood harmonic ratios using vibrating strings, is it possible instead to understand harmony based on the "vibrations" of the moon, Jupiter, and Saturn? The numbers 120, 128, and 135, and their disposition within the array in table 1.1 (of the powers of 2, 3, and 5) might have caused megalithic minds to fill in the gaps within that array. If one looks for the largest number under Jupiter (equal to 135), one can successively double 5 to become 80 (less than 135), and a similar process fills in the blanks in table 1.2.

TABLE 1.2

5	80	Lunar year (120)	90	Jupiter (135)
0	Saturn (128)	96	72	108
0	3	9	27	

Looking at the bottom row, $128 \times 3 = 384$ which is greater than 135. Then divided by 2, equals 192, and divided by 2 again equals 96, so that

1. $128 \times \frac{3}{4} = \textbf{96}$. And moving right,
2. $96 \times \frac{3}{4} = \textbf{72}$, and
3. $72 \times \frac{3}{2}* = \textbf{108}$.

Moving upward from Saturn = 128

4. $128 \times \frac{5}{8} = \textbf{80}$. Moving right to 96,
5. $96 \times \frac{5}{4}^† = \textbf{120}$, then to,
6. $72 \times \frac{5}{4} = \textbf{90}$, and
7. $108 \times \frac{5}{4} = \textbf{135}$.

This means that

- moving right (by increasing the power of 3), the limit 135 forces us to halve a result twice, most of the time, while
- moving upward (by increasing the power of 5) the limit 135 forces us to halve the result three times, on occasion.

These operations within the table are the harmonic intervals between definite vibrations or tones, many of which appear unoccupied in the

---

*The aim is to always stay within the octave and so multiplication or division by 2 can always achieve a number within the octave. There is no fixed rule for 2 but the single 3 in the denominator is essential, and the tone numbers generated must be within the octave—a law of limits.

†The interval ratio 5/4 is smaller than 3/2, so fewer corrections are required when traveling upward in these tables through still applying the octave limits for tone numbers.

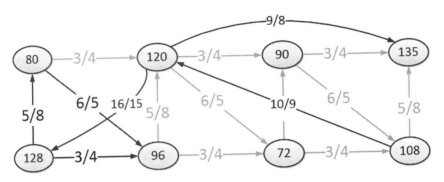

Figure 1.3. The musical intervals latent within table 1.2.

synodic solar system seen from Earth. Today we would say that the rows are a series of fifths (3/2) in tuning order and columns are major thirds (5/4):

- 3/4 is 3/2 (the fifth), which we know as the largest interval within an octave. Its denominator is sometimes 4 to reset a tone back to being within the limit. Such changes involving powers of 2 never affect location within octaves but affect changes between adjacent octaves.
- 5/8 is 5/4 (the third), which we know as the major third, another large interval within the octave.

There are other relationships in this table,

1. The diagonal from 80 to 96 is 6/5, the minor third;
2. 80/72 is 10/9, which is a shorter tone than 9/8;
3. 90/80 is 9/8, the longer tone;
4. 128/120 is 16/15, the semitone.

These last three intervals are those required by modal music to make musical scales, and scales are the ancient form of our own melodic music, though here, they are being revealed without a monochord being struck or a human ear ever hearing a sound! The same rule applies throughout this table for

- 96/90 and (72 × 2)/135 are both semitones of 16/15,
- (72 × 2)/128 and 108/96 are both longer tones of 9/8.

That is, vectors of the same angle and length within this state space of 3 and 5 traverse the same interval between tones, irrespective of where they start. This is rather like a game of chess with its moves, and indeed the "knight's move" proves to be crucial as 128/90 = 64/45 = 1.42, an approximation to the square root of 2, called the "tritone" in harmony, being the interval between tones that is least harmonious with respect to another while standing halfway within a musical octave interval. This is the famously secret or feared irrational square root of 2 of the Pythagoreans, cropping up time and again in later chapters and many different cultures.

One notices from figure 1.3 that the three large harmonic intervals—3/4, 5/4 (as 5/8), and 6/5—are reminiscent of the relationship between the sides of a 3-4-5 triangle, the smallest triangle possible in which the sides are all whole numbers. This seems a natural consequence of the fact that the primes 2, 3, and 5 are all in relationship in the side lengths 3, 4, and 5, which, being Pythagorean, are related through their squares as 9 + 16 = 25 so that a square of side 5 containing 25 squares can have 4 × 4 = 16 with 9 left over or 3 × 3 = 9 with 16 left over. This creates an interesting diagram of the square shown in figure 1.4 in which all the useable modal intervals that define musical scales can be shown as within the square of 25. The square includes one not yet seen here, 24/25, the chromatic semitone.

Intervals with integer numerator and denominator clearly interact within rectangles and squares to form a class of (what we call) harmonious intervals based on the primes 2, 3, and 5 and not employing 7, 11, or later prime numbers. This diagram presents an ideal encapsulation of knowledge about the early numbers, the square being easy to share. This appears borne out by the 1735 diagram of paintings on a Spanish rock shelter at Cachao-da-Rapa (from the Copper Age), in which many subdivided squares and rectangles including squares with side 5 were presented, one of these including a visual proof of Pythagoras's theorem for the 3-4-5 triangle.

1	2/1	3/2	4/3	5/4
2/1	4/3	6/5	8/7	10/9
3/2	6/5	9/8	12/11	15/14
4/3	8/7	12/11	16/15	20/19
5/4	10/9	15/14	20/19	25/24

The Square of Generation

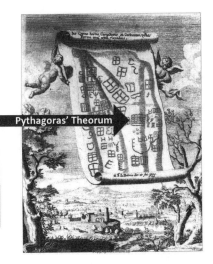

Pythagoras' Theorum

Figure 1.4. The generative nature of the square of side five (*left*). All the ratios used within ancient music based on number can be generated within this square. If the small squares are taken to be a unit = 1, then the development of square area has interesting generative properties as multiple squares form rectangles and squares. In the square of 4 squares the fourth square is in the ratio 4/3 (called the fourth), of 9 squares the ninth is 9/8 (Pythagorean tone), the sixteenth is 16/15 (just semitone), and the twenty-fifth is 25/24 (chromatic semitone). With rectangles, other ratios evolve such as 2/1 (octave), 3/2 (fifth), 4/3 (fourth), and 5/4 (major third). The second row (and by symmetry second column) generates 6/5 (minor third) and 10/9 (just tone). The 4-by-5 rectangles implicit in a five-sided square have a diagonal length of 5, enabling the Copper Age diagram (*right*) to present a visual proof of Pythagoras's theorem millennia prior to Pythagoras.

Returning to table 1.2 for the limit 135, one notices that numbers less than 135 must never be less than half of 135 if they are within the octave below the limit. Such tables will always therefore only contain *all* of the regular numbers less than the limit yet more than half the limit. They are therefore, by default, the tone numbers that would naturally exist in an octave with high do equal to the limiting number. From this we also note that 135 cannot form a rational octave because low do would be equal to 135/2 = 67.5. If we make the limit 90 × 2 = 180, we can present all of the same numbers since 120, 128, and 135 are all within the octave less than 180.

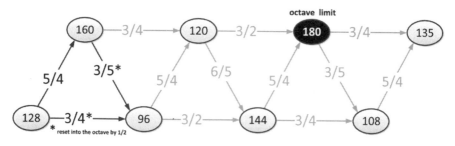

Figure 1.5. The skewing of the powers of 5 relative to the powers of 3.

Another small modification to such tables is also indicated, because the upper register of numbers relates strongly to both the number directly below each number and to the number below but to the right, so that 160 relates to both 128 (5/4) and 96 (3/5). It is also true that, as higher limiting numbers form larger tables, there will always be fewer numbers in the higher row than in the row below, because five is larger than three. As a result, it appears that the ancient world shifted each higher register half a step to the right, as in figure 1.5. This caused the full extent of such a table to look like a mountain of regular numbers (see figure 1.6 for limit 720.)

THE BUILDING OF HOLY MOUNTAINS

The lunar year's relationship to the Jupiter and Saturn synods is best presented within a skew matrix of powers of 5 versus powers of 3. Ernest McClain probably came to this format when decoding Plato's tuning texts, and through recognizing that mountain topographies were commonly encountered within ancient texts, leading to McClain's primary technique for interpreting ancient texts, by building the harmonic or "holy" mountains associated with the numerical references found within the texts.

The simplest such mountain is that of limiting number 60, introduced in chapter 2 as having much to say about the harmonic gods of Mesopotamia. The second simplest mountain is the one based on 180, the age at which Isaac in the Bible dies, his mother being 90 when she births him. In the Bible (see chapter 3) the patriarchs grew by doubling through 360 to 720, at which point the ideal forms of holy mountains can be seen. Figure 1.6 was generated by the web-page application Harmonic Explorer,* written for Ernest McClain, who (in 2012) was still calculating large mountains by hand.

Ignoring the duplicates for E and C, 720 generates the twelve tone-classes of the chromatic scale, each separated by semitones, alluded to by the twelve children of Israel of the patriarchal line. The lunar year is then valued at 480 units or 40 units of time per lunar month, while the Saturn synod has become 512 and the synod of Jupiter 540. At the root of 720, below 360, 180, and 90 lies 45, whose factors are 3 × 3 = 9 and 5. This root has 9 elevated by 5 so that ADM = 1, 4, and 40 (the Hebrew Letters for Adam—the second a is not a "letter"—are ADM, aleph=1, dalet=4, and mem=40, so that ADM is 1 + 4 + 40 = 45) is at the root of 720, while if 720 is doubled, Adam appears again as 1,440 but now in decimal-place notation, a limit that, as we shall see in later chapters, starts to unlock other astronomical synods and calendrical periods.[4]

One important calendrical period can already be seen in Adam's patriarchal mountain, since 720 divided by 40 units is 18 lunar months. This is not a time period familiar to what we know of the ancient world, but the Olmec of Mexico (1200 BCE onward) appear to have used it as their supplemental glyphs tagged onto their Long Count inscriptions. Doubling of 720 to 1,440 and then to 2,880 at last completes the musical or harmonic part of the mountain, its darkened bricks forming what McClain thought was "Ishtar's bed," and it is then that the Olmec report the arrival of the eclipse year onto the mountain, as part of their feathered serpent Quetzalcoatl (also a Mexican name for Venus) around which a supra-harmonic astronomical cult, shared by the Maya and Aztec, was formed (see chapter 7).

The mountain for 720 has other resonances to the megalithic culture that could have evolved such a system from astronomical metrology rather than acoustic calculations. There were 40 megalithic inches within a megalithic yard and, when used for counting months rather than days,

---

*harmonicexplorer.org

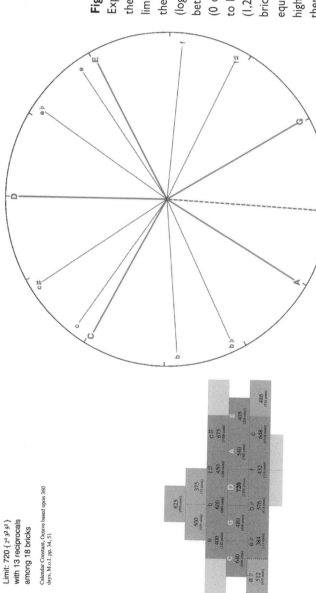

Limit: 720 {2⁴ 3² 5¹}
with 13 reciprocals
among 18 bricks

Calendar Constant, Octave based upon 360
days, M.o.I. pp. 34, 51

Figure 1.6. Harmonic Explorer view of (*left*) the holy mountain for limit 720 and (*right*) the tone circle (logarithm base 2) between low D = 360 (0 cents*), clockwise to high D = 720 (1,200 cents). The darker bricks are the tones equally reachable from high and low D and therefore symmetrical on the tone circle.

---

*Cents are a modern system of dividing the tone circle into 1,200 equiangular parts (100 parts, or cents, for each semitone) so that tones can be quantified more accurately within their naturally logarithmic domain where the same interval has the same number of cents, helpful within the microtonal environment of these mountains (see appendix 2).

18 such yards (the supplemental glyph number) have 720 inches. This enabled a model of this harmonic limit to be built using 18 megalithic yards in which the lunar year of 12 months would be 480 inches long. This is similar to what has been found in the road structure between the Sun pyramid and that of Quetzalcoatl at Teotihuacan (see page 178). Saburo Sugiyama has methodically recovered the length of the megalithic yard as being the key Teotihuacan measurement unit (TMU), and 1,440 TMU (megalithic yards) define the distance between the centers of these two pyramids. The same model, in different units, is to be found in the Parthenon of Athens (see chapter 4)—only there, Athena replaces Adam.

One can therefore reconnect (*a*) Ernest McClain's work on ancient harmony and his interpretation of ancient harmonic clues to (*b*) the harmony of synodic astronomy and the megalithic astronomers who measured synods using metrological methods. This makes a more compelling case for ancient tuning theory as having arisen *from the heavens.* Musical tuning theory became associated with religion for reasons greater than being used as a convenient acoustical metaphor. Harmony was obviously considered a well-developed cosmological model of the world and its process.

2

Heroic Gods
of the Tritone

In chapter 1 we saw that the megalithic astronomers did not need a fully externalized notation nor a procedural arithmetic to recognize, appreciate, and calculate harmonic ratios between the metrological lengths formed using day-inch counting. That same metrology, as it developed in the ancient Near East and later cultures, came to be called "historical metrology" and can be found scattered over the Old World in field sizes, public spaces, and buildings. While acoustic string lengths remained metrological, when measured, numerical tuning theory embraced the alternative *arithmetical* developments of the Sumerian and succeeding Near Eastern cultures. A harmonic mountain only required multiplication by primes 2 (doubling by "adding to itself"), 3 ("adding a third time"), and 5 (doubling twice then "adding again"). Kings, priests, and cities could afford scribal computors, so calculation (*a*) took over from geometrical methods and (*b*) such calculating could investigate larger limiting numbers. It would have seemed natural to conflate the actual, astronomical heavens with a new harmonic heaven in which planetary and other gods lived, a heaven that rises up through human calculation, in search of where these gods reside, as the synodic planetary periods but also as new heroic gods made of larger numbers, who seemed to have forged the world of harmony and, in some cases, helped mankind to see it.

Ancient civilizations had a propensity for building high pyramids and ziggurats. Pyramids can be viewed in two ways and sometimes both ways at once: as symbolic of the mountain of heaven, or of the geodetic earth itself, that is, as a "primordial hill."* The most famous ziggurat, that of heroic god Marduk, was geodetic in terms of representing the shrinking parallels of latitude in the Northern Hemisphere while also being a harmonic monument to his distinctive limiting number, 8,640,000,000—a number shared by the flood heroes and time periods of other traditions.[1] The pyramidal form of a building could therefore be used to represent the harmonic heavens and the geodetic earth. When harmonic its size could embody the very large harmonic numbers appearing in literary references to harmony, since these are best presented as a two-dimensional tuning matrix or "holy mountain."

HARMONIC CALCULATION AS PYRAMIDAL MOUNTAINS

To review: Successive multiplicative tuning intervals *appear* to run along the rows and columns of harmonic mountains, though the mountain is *actually* a matrix of powers limited by a number. The horizontal dimension represents tone numbers involving successive powers of prime number 3, while the vertical dimension represents successive powers of prime number 5. These intervals all emanate from the (bottom left) "cornerstone" harmonic root of 1, that is unity—the matrix unit, where primes 2, 3, and 5, are all to the power zero. The zeroth power of any number equals one, the *common unit* on which all the harmonic numbers in the matrix are based.

When all of these powers of 5 and 3 are translated into a given octave, the limiting number given to the octave chops off three sides of an extensive mountain (see figure 3.3 for limit 720), beyond which rationality breaks down, because the factors available to a given limiting number run out of either 2 (to the right), 3 (to the left), or 5 (below the

---

*A term from the Egyptian Memphite theology, found for instance in the "Hymn to Ptah-Tatanen." See Naydler, *Temple of the Cosmos*, 37.

base), causing the breakdown of rationality, because all fractional results fall by definition outside of the mountain. Put another way, the location of a limiting number on a mountain is defined by its factors of 3 and 5, these placing it relative to the cornerstone $= 1 \times 2^n$. For example, the limit of Sumer's "father of the gods" Anu was 60, leading to a matrix that will be three rows high because

(a) below the matrix there are *negative* powers of 5 and these are never integers;

(b) the zero power of 5 forms the bottom row;

(c) the first power of 5 equals 5, which is less than 60, initiating the second row;

(d) the second power of 5 equals 25, less than 60, forming the third row; and

(e) the third power of 5 equals 125, greater than 60 and hence: there is no fourth row.

Note well the following: The numbers within mountains are called tone numbers. These are then pitches (frequencies) despite many ancient scales being defined as descending in pitch with the lengthening of string length. The allocation of note classes in McClain's mountains rose with tone number and so must indicate pitch on his mountains.* Later chapters (especially chapter 6) show that reciprocal tones relate to the ability of each modal scale to also express another "twin" mode; an ascending scale being also, in descending, another twinned scale, within any given octave.

The reader should also note that megalithic counting measured the number of days or months within astronomical time periods, and so they were *tone* numbers, that is, frequencies. This perhaps explains why Ernest McClain used tone numbers: the origins of ancient tuning theory and its mountains were astronomical frequencies rather than the result of musical experimentation.

---

*See my discussion of the "rusty compass" in appendix 2, "Ancient Use of Tone Circles."

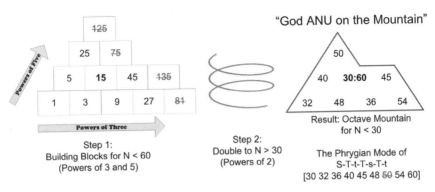

Figure 2.1. The ancient Near Eastern model of harmonic calculation leading to holy mountains. See box for a detailed explanation.

Steps Forming the Harmonic Mountain for Anu = 60

When powers of 3 within 60 are considered, the powers of 5 will be making tone numbers on a succeeding row five times larger than those beneath it. Since 5 is the largest prime involved in harmony, the powers of 3 not multiplied by 5 will be

| 3   9   27   8̶1̶

where 81 is too large to belong to the limit of 60, and the second row will be

5   I5   45   1̶3̶5̶

while the third row will be

25   7̶5̶.

This matrix of harmonic roots will then be

25   7̶5̶
5   I5   45   1̶3̶5̶
| 3   9   27   8̶1̶.

One can see that to the left the powers of 3 will be less than I and numbers will become fractional. To the right, the extent of whole numbers included in the limit of 60 is limited by the highest power of 3 less than the limit. There are four powers of 3 (namely,

0, 1, 2, 3) below 60, hence four on the bottom row. When multiplied by 5 the next row only has three powers of 3 less than 60 (0, 1, 2) and the second power of 5, 25, allows no power of three except the number 1, the zeroth power of 3.

But the octave limit 60 is shown 30:60, and this means that all those factors of 3 and 5 not already within that 30:60 octave will have to be doubled until they are. The cornerstone must be doubled to the *only* power of 2 in the series 1–2–4–8–16–32 . . . that can exist within the octave, in this limit 32, the fifth power of 2. The case is similar with 3–6–12–24–48 and so forth, leaving a new set of numbers, now products of 2, 3, and 5, as an octave populating the octave of 60.

```
50
40      30:60   45
32      48      36      54
```

The tone numbers in the rows now correspond to a tuning order* where the adjacent number pairs all *relate* as fifths ascending toward 60 from left to right, and as fourths (4/3) descending toward 30 from right to left. The small limit of 60 cannot even afford a five-tone pentatonic scale made up of successive fifths for practical music. But the mountain is starting to evolve major and minor thirds above and below the limit due to the action of prime 5. The number 60 was a perfect limit to demonstrate what is possible with such mountains, where the note D is taken to bound the note classes found within an octave, and this avoids the brutal approach of trial and error in studying the harmonic domain, allowing the lowest possible number to be known for a given harmonic phenomenon to be realized.

---

*The "tuning order" of fifths is the order in which a spiral of fifths is established when practically tuning a harp (see figure 5.4). As already stated, this happens to be true because each number is multiplied by 3/2, raising its powers of 3 as a factor. Note that forcing all tone numbers into the octave range 30:60 automatically resets tone numbers into the octave—a task done by ear when manually tuning strings.

We should expect D as 30:60 to be raised up one row, and, because this row's numbers can only contain a single factor of 5, the limit 60 only affords two intervals to D of 30:40::45:60. When this set of tone numbers is divided by 5, 60 is seen to have raised up the simplest numerical form, 6:8::9:12. The octave 30:60 has four notes below, on "the earth," 30:32:36:48:54:60, and only two of these (48 and 36) form the major and minor thirds 4:5 and 5:6 in relation to 30, showing that the skewing successive rows developed by McClain were crucial for his interpretation of ancient texts to correspond to their harmonic allusions. The rows of the tone mountains then overlap like the bricks in a wall.

Figure 2.2 reveals two important features of such mountains of harmonic numbers. First, is the core process within the skew matrix,

- of intervals 3:2 left to the right
- of intervals 5:4 at +45 degrees
- of intervals 6:5 at +135 degrees.

Second, the form emerging is of a pyramid of brick-like numbers, each layer raised up by multiplication by 5. This is clearly seen between

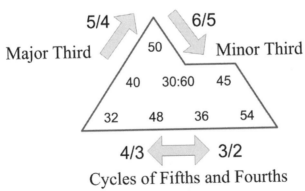

Directions on the Mountain Are Musical Intervals

Figure 2.2. The three major tone directions within the mountain: horizontal as fifths and fourths, ascending right as major thirds, and descending right as minor thirds. This framework for larger limits comes later to define the tones of just intonation and its seven modal scales.

40 and 50, harmonically 5:4, the major third, and between 50 and 60, 6:5, the minor third.

THE SUMERIAN REVOLUTION

Such holy mountains are thought responsible for the Sumerians' ability to comprehend harmony as an extensive yet law-conformable domain in two dimensions. To codify important parts of their scheme, they saw 60 as their primordial god Anu. Within his octave one can see 50 atop the mountain, the number similarly associated with Enlil, king of the gods. The third in this trinity was 40, Ea-Enki, the god of the waters. The "waters" were the chaotic fractional world within which a limit is set and harmony established through the computational steps of the mountain; of 5:4 "to step up" and 3:2 "to step forth." We can see Ea stepping up in the cylinder seal of figure 2.4 and Shamash (Saturn/Sun) stepping up in figure 2.3.

Figure 2.3. Shamash steps up as if at the angle of the 3-4-5 triangle, expressing mountainous harmonic landscapes since the hypotenuse is five units if the base is three units. Shamash has joint meaning of the sun and planet Saturn, who is the cornerstone, then stepping up by 15/16 (a semitone) to the lunar year, located as Anu = 60.

The sexagesimal base of 60 gives birth to the three primes 4 (as 2 × 2), 3, and 5 since 3 × 4 × 5 = 60 (with lower D = 3 × 2 × 5 = 30). These are also the side lengths of the first Pythagorean 3-4-5 triangle showing how deeply harmony is rooted within the world of numbers and geometry. Such mountains are the powers of prime 3 and prime 5 below a given limiting number, wound up by powers of prime 2 to enter the octave of the limit. The three primes are represented at every point of the mountain by the three most harmonious intervals: 2 as the eighth (or octave note), 3 as the fifth (note), and 5 as the third (note) of the octave, the numbers referring to the classic order of notes within musical scales.

Such mountains were as fundamental a discovery for the ancient world as algebra and theorems were for the Greeks. They replaced specific calculations of musical scales with an abstract extendable domain that could be studied in its own right. Anu, Enlil, and Ea-Enki were later overtaken in myth and, in parallel, harmonically by a range of gods associated with the most important aspects of the new domain. The later gods were found within mountains of various new limits. One can adapt Samuel Kramer's summing up of Sumerian creation concepts as being a harmonic myth of emergence:

1. First was the primeval sea Nammu. Nothing is said of its origin or birth, and it is not unlikely that the Sumerians conceived of it as having existed eternally (like the field of numbers).

2. The primeval sea begot An (the cosmic mountain of limit 60) consisting of heaven and earth united.

3. Conceived as gods in human form, An (heaven) was male (3 and 5), and Ki (earth) was female (2). From their union was begotten the air god Enlil (50 = 25 × 2).

4. Enlil separated heaven from earth, and while his father, An, carried off heaven (the balanced powers of 3 × 4 × 5 = 60), Enlil carried off his mother, Ki. The union of Enlil (50) and his mother, Ki, set the stage for the organization of the universe, the creation of the moon and sun. Man was created by EnKi (40), the water god, son of Enlil and Ki as 5 × 8 = 40.[2]

In terms of Anu's mountain, Enlil (= 50) is at the top having carried off Ki who births EnKi (= 40), god of the harmonic waters. An has carried off heaven in a slightly different, balanced direction, to become Anu (60), the first power of 60, the path of the middle way. It is worth noting how the stars of ancient Mesopotamia were similarly described. The northern stars were in "the path of Enlil," the zodiacal stars in "the path of Anu," and the southern stars in "the path of Enki." This would have formed a seamless astro-harmonic vision of the creation not dissimilar to our own solar usage of equatorial stars and those beyond the Tropics of Cancer and Capricorn.

Early musicological thinking about numerical scales in the ancient world should perhaps be reviewed as originating within McClain's harmonic mountains, especially as to its first two rows, since this can intellectually resolve the seven-note diatonic* scale that has semitones (16/15) more harmonious than the Pythagorean semitones (256/243) populating an octave (by ear) with successive fifths and fourths. When a third row is included, and the factor 25, one can see all of Enki's creation of five modal scales in the darkened bricks of figure 3.3 for limit 720 (see also chapter 6 on scales), the modes inherent to just intonation (the form of tuning represented by such mountains) made up of a *mixture* of fifths and fourths. The first three rows of larger mountains such as 720 therefore defined a *modal* music of different scales (chapter 6) in which instrumental pure tones are available around D. These scales cannot be modulated as with modern instrument tuning (that is, using equal temperament), so that harmonic mountains were useful to ancient musicalogical and cosmological thinking yet these today appear overly theoretical.

Any vector on the mountain will involve a differential ratio of 2s, 3s, and 5s that represents a constant harmonic interval and, as already stated, the same vector will apply throughout any such mountain for a given D, to the same effect. This was clear with the vector "one step right" always equaling the fifth and the "one step upward (skew)" always

---

*A "diatonic" scale is one of five whole steps (whole tones) and two half steps (semitones) in each octave.

equaling a third. But that is just typical of this matrix approach, as is seen later in the tone 9/8, separating two fifths, which is "two steps to the right," found between the lunar year and Jupiter synod but also anywhere else, similarly separated.

So where is Saturn? To see Saturn one needs to see that the cornerstone (= 32) is 16/15 of D = 30, one step to the upward diagonal, but that the Jupiter synod as 9/8 lunar years cannot be seen beneath the limit of 60. The solution is to move D to the right by multiplying the limit of 60 by three to define the limit of 180. This limit was proposed in chapter 1 as having been ideal for viewing both Saturn and Jupiter synods astride D = 180, as G and A. The cylinder seal of figure 2.4 appears to show how the creation moved on from the limit of Anu (= 60) to a limit of at least 180, as found between Saturn, Jupiter and the lunar year in chapter 1.

Figure 2.4. Mesopotamian deities, low relief: Shamash, the sun god, rising in the morning from the eastern mountains holding a long knife to cut his way out of the earth between (left) Ishtar (Sumerian: Inanna), the goddess of the moon/morning star, and (right) Ea (Sumerian: Enki), god of the waters, with (far right) his vizier, the two-faced Usmu. If the hillocks represent lunar year on left and Jupiter on right, Shamash is also appearing where Adam and Athena will, as the 18-month "supplemental period" of the Mayan Long Counts. Note also the similarity of the knife of Shamash and that of Quetzalcoatl in figure 8.10 on page 185.

Ishtar (Venus) was thought to be the daughter of the moon god Sin, whose number was 30, the lower bound of Anu's octave limit 60. Ishtar's given number in the pantheon below 60 was 15, the harmonic root of that "location." This affiliation of Venus with the moon is presented in the sacred image of the star and crescent (for example, in the flags of Islamic countries), where she emerges from the moon. Her planetary position in the harmonic matrix (see chapter 7, under "Osiris as Resurrected Fertility God") finds her root (15) being raised to the fourth power, locating her limit of 60 to the power of four or 12,960,000.* Looking at figure 2.4, in relation to the holy mountain limit 720, we see Ishtar's winged form above the lunar year, while Ea-Enki is stepping up to the location of the synod of Jupiter. This shows the strong control of the right-hand Jupiter hillock over the lunar year's hillock, an interval of 9:8. Between the two is Shamash rising between these two "hills." The two-faced advisor Usmu symbolizes the symmetry found in the tone circle of any octave and its twin tone numbers and (see chapter 6) twinned modal scales. Between the ascending fourth of G and the descending fourth of A, within the World Soul's tone circle, Shamash emerges as the tritone located in the *tonal region* between the lunar year (G) and Jupiter (A) synod; this is probably why Saturn was conflated in the name Shamash with the sun, being the "sun of the night."[3]

In the development of the World Soul ($180 \times 4 = 720$), Saturn (512) is uniquely the tritone a♭† to D, yet also the cornerstone that defines the whole planetary matrix. This comes about because the cornerstone has the tritone vector from D on the mountain for 720. Located between G and A in the *tuning order,* it then becomes possible to access the tritone and weave the seven modal scales of harmony (see chapters 5 and 6). D = 720 is the smallest limit for this to be true: far enough away from the cornerstone in its primes of 3 and 5, that the tritone then appears between G and A *on the tone circle,* causing all seven scales to become possible. Modal scales arise from prime 5 (Plato's humanly divine number) and, in later literature, scales are alluded to as tribes, five or

---

*This limit is the same as Plato's sovereign number, surely no coincidence.

†Lowercase notes indicate just intonation.

seven in number. By the sixth century BCE, the compilers of the Bible in Babylon turned this special situation (of the cornerstone also being tritone) into Adam = 45, while for the Greeks, the Parthenon would symbolize the same powerful truth: that the World Soul was harmonious but also creative and destructive, for the human world. Such shocking simplifications can only have emerged from considerable efforts to envisage the whole domain of harmony when opened up by the three primes seen within harmonic mountains or pyramids.

A HARMONIC SPACE RACE

Harmonic knowledge would have been a deeper way to see into the world of the gods, informing the development of sacred calendars, of astronomical periods and events. When kings merged their identity with the gods to become god-kings, most famously in Egypt but elsewhere as city or tribal deity-rulers, tone mountains offered a rich source of sacred iconography alongside the fertile ground of harmonic allusions possible within sacred narratives and myths. Until rather recently, we were without this harmonic key to many of these ancient and inevitably cryptic harmonic symbols.

In the 1970s a school of harmonists appeared with the proposal that ancient spiritual artifacts had a strong component of harmonic coding. Ernest McClain's groundbreaking book, *The Myth of Invariance* (1976), came only a few years after *Hamlet's Mill: An Essay Investigating the Origins of Human Knowledge and Its Transmission through Myth* (1969) by Giorgio de Santillana and Hertha von Dechend—a book similarly proposing that astronomy and numbers were invariants to be found within ancient myths and designs; both books appearing in paperback in 1977.

The greatest civilizations seem to have gone further and created some gods that were representations of purely harmonic realities, gods who often fought earlier gods and demons, defeating or sometimes killing them. One such was Marduk, god of the Semitic Babylonians, and similar to the later Greek Zeus in the events that created him and related in some way (as we shall see) to Jupiter. Anu had to die before

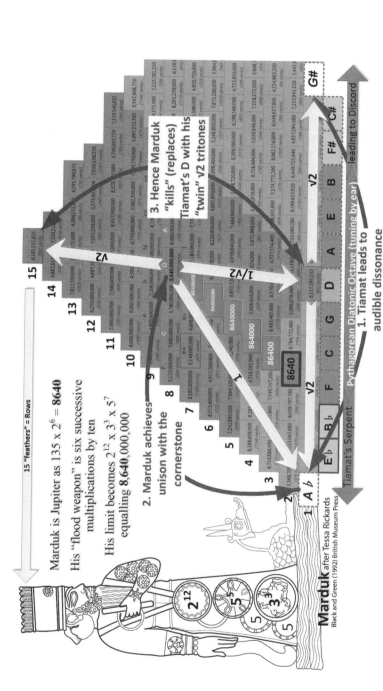

Figure 2.5. Iconography of the god Marduk, full of numerical disks symbolizing his victory over Tiamat, as a very large number. Tiamat's crimes against harmony include making the cornerstone A♭ audibly less than G♯ so creating dissonance. Marduk (8,640) uses his flood weapon until his D is in unison with A♭ because he forms a tritone (√2) to Tiamat, while his top brick on the fifteenth row is then reciprocal twin to Tiamat. The act is presented as revolutionary in leading to just intonation, which avoids giant numbers.

Marduk could be born, whereupon he exceeded his father, Enki, and became a new type of super-god who cleared out some noisome lesser gods allied to Tiamat, a primordial harmonic entity. He killed Tiamat for her abject passivity to the often discordant tones generated during a rebellion by lesser gods, some of whom had created humankind to do their former work. The Hebrew god YHWH had similar problems with giants: gods slept with the daughters of men who then gave birth to a giant race, requiring a flood to eliminate them. It was from Marduk's "flood weapon" story that the biblical flood story of Noah was taken, and flood stories are common to many ancient cultures and often involve, McClain believed, giant numbers, all of which are dispatched by means of a flood: super-god Marduk responds by killing Tiamat, YHWH causes a biblical flood (chapter 3), and Indra in the Rg Veda (see page 140) kills the serpent Vrtra.

Marduk (8,640,000,000) had seven powers of 5 while YHWH (777,600,000) only five powers of 5 but centuries of harmonic thinking separated the arising of these two super-gods of harmony, and YHWH had new harmonic lessons (see chapter 3). We have seen that the prime number 5 ($\times$ 2 = 10) is a perfect partner for mitigating those problems associated with only using prime 3 to populate the octave with fifths, in the Pythagorean style (the python being a *serpentine* cycle of fifths). Seven Pythagorean tones can make a seven-note or heptatonic scale, but the twelve note classes familiar today can, formed from a cycling of fifths, make a monster whose last tones are enharmonic approximations to the perfect tritone of G#, which should equal A♭, but audibly didn't for serpentine tunings. When one looks at the numerical reality of such a Pythagorean tuning system, very large powers of 3 accumulate in the repeated fifths, making tone numbers difficult to calculate or render into practical string lengths as whole numbers. Hence, seen from the perspective of numerical tuning theory, Marduk and other flood heroes provide a critique of only using the powers of 3. Marduk on his mountain can pump himself up with excessive 5 power to create a flood of numbers fifteen layers deep. But to what end?

It's all about the tritone: if Tiamat is laid along the bottom, a serpent of fifths along the "earth" of Marduk's enormous mountain, then

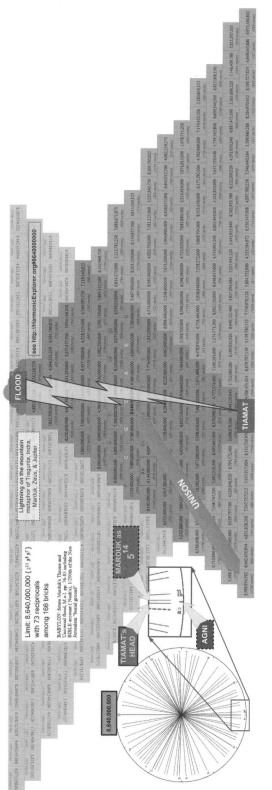

Figure 2.6. The harmonic mountain for Marduk's number, shared by Indra, Brahma, and other super-gods. With 21 along the base and 15 high, the peak stands above 729.

she has twenty-one Pythagorean tones in a world of cacophony. He (having the same number as Vedic Indra for his throne, 8,640,000,000) has made seven decimal zeros of his seven powers of 5 by inflating them with twelve doublings. But his flood weapon then appears at the top of his ziggurat, which is *fourteen* powers of 5 "tall," towering above Tiamat below. The top of his mountain casts a shadow of (darkened) symmetrical twins, all the way down to the sixth power of 3 (729—Plato's tyrant number), the top and bottom generate the most perfect tritone possible using the powers of 3 and 5. The mountain brings every location on it into the same octave and by then dividing them many times by 2, a modern fractional number between 1 and 2 emerges. Marduk's peak is then 1.42108, or 598 cents. Its shadow on Tiamat is 1.42383, or 602 cents. A perfect tritone would be 600 cents, the harmonic square root of 2, but that is irrational and can only be approximated. The purpose of the exercise seems to be to kill Tiamat and then "kipper" her (figure 2.7), placing part of her former self above another part, and this is exactly what is done when a single power of 5 is used to create a row above the bottom row of a mountain. Diatonic scales of seven notes then become possible using a much smaller limiting number, a scale with better semitones than the heptatonic. The similar thing done by Indra cuts the serpent Vrtra twice, with the head portion left on the earth and the tail in the sky with the middle (with D) central on the first row, raised by 5. This is the full format of just intonation, in which all the seven modes or "tribes" can appear and gain their full freedom to mutate into each other via the tritones (see chapter 5).

The displacement of part of Tiamat by 5/4 comes by including harmonic numbers from the number field that contain a single factor of prime 5, the same principle that creates holy mountains, and Marduk is dramatizing that harmonic lesson. If $3^4 = 81$, the fifth segment of Tiamat is "cut," then the numbers up to 3^3 remain on earth and what were numbers 3^4 and higher are reset to being 5×3^0 or higher. This resolves the root of the moon (G = 15 = 5×3^1) and Jupiter (A = 135 = 5×3^3), leaving Saturn ($a\flat = 2^n \times 3^0 \times 5^0$) as cornerstone. At this point the vector nature of the harmonic mountain is revealed as responsible for forming the modal scales (see chapter 6), since there are

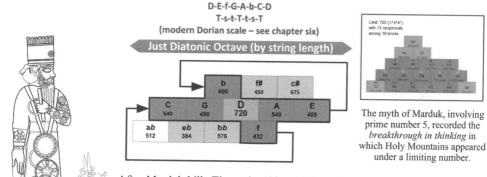

D-E-f-G-A-b-C-D
T-s-t-T-t-s-T
(modern Dorian scale – see chapter six)

Just Diatonic Octave (by string length)

The myth of Marduk, involving prime number 5, recorded the *breakthrough in thinking* in which Holy Mountains appeared under a limiting number.

After Marduk kills Tiamat he "kippers" her, placing sections of her former octave above others; exactly what is done by numerical tuning theory when a single power of 5 is used to create a row above her "serpentine" row of fifths, and become a "mountain" of more flexible "just" intervals.

Marduk after Tessa Rickards
Black and Green (1992) British Museum Press

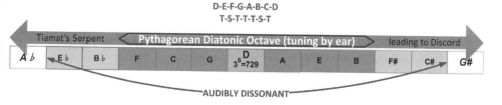

D-E-F-G-A-B-C-D
T-S-T-T-T-S-T

Figure 2.7. The cutting up of the disharmonious monster Tiamat so as to place her central heart of D above her front section using prime 5.

three vectors throughout these mountains that form the three intervals of which just-tuned scales are made: 9/8, 10/9, 16/15.

If D = 45, then Saturn as $1 \times 2^5 = 32$ gives the tritone of 45/32 = 1.4063, while the Saturn to lunar year (30) interval is 16:15 = 1.0667, the just semitone. We have already resolved lunar year to Jupiter as 135:120 = 1.125, and the location in between is now D = 720 while it was Tiamat's D = 729, now excluded by the limit. The central part of Tiamat became a raised pentatonic, which was chosen by the biblical harmonists for their patriarchs.

We are now ready to contemplate Marduk's ziggurat as a harmonic monument to his victory (figure 2.8). If the Tower of Babel of the Bible's book of Genesis, it would have been contemporaneous with the compilation of the Bible in sixth-century-BCE Babylon. Like the mountain for Marduk's number (figure 2.8, right) it has 15 levels, judging by

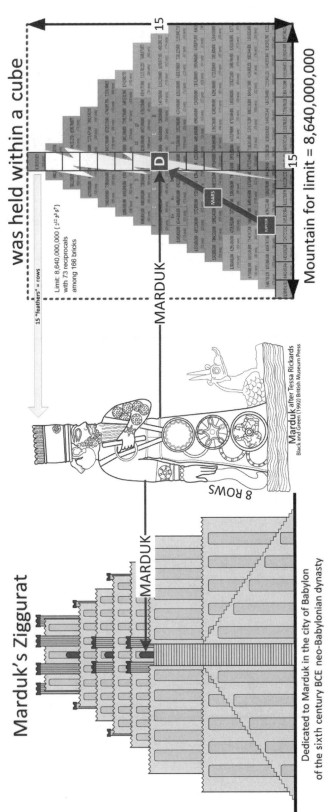

Figure 2.8. Etemenanki (Sumerian for "temple of the foundation of heaven and earth") was a ziggurat (*left*) dedicated to Marduk (*center*) in the Babylon of the sixth century BCE (neo-Babylonian dynasty). Originally 91 meters in height, little remains of it now except ruins. Etemenanki is considered the possible inspiration for the biblical story of the Tower of Babel. The holy mountain for flood-hero number 8,640,000,000 (*right*) can be compared to the ziggurat as having 15 levels, with correspondences noted between Marduk's iconography, flood mountain, and ziggurat.

the four single floors and other blocks strongly implying a metric of one-fifteenth of the total height and width. Marduk's throne would be on row 8, the center of its cubic volume where the steps lead to a door. The side length of the cube formed 4,320 *shu-si,* the Sumerian digits, of which there were 30 in an Assyrian foot and 60 in a double-foot cubit, making the 4,320-digit side length 180 Assyrian cubits. This indicates tuning theory was informing Marduk's monument since 4,320 is half of Marduk's "head number"* of 8,640.

---

*A "head number" is the part left over when decimal zeros are removed from a number, the head then containing the residue of powers of primes 3 and 2. We include one zero in the head number because Jupiter, when 8,640, is 720 × 12.

3

YHWH Rejects the Gods

The Bible starts with Genesis, a book that sets the scene through the emergence of the first man, Adam. The number-letter equivalence of his name in Hebrew gematria sums to the lowest starting point for just intonation and hence, arguably, makes his name a reference to the planetary matrix. First-millennium phonetic alphabets had numbers associated with each letter so that numbers could be recorded. This allowed numbers to be written as words, enabling harmonic numbers to be hidden within names—remembering that Adam was given the power of naming by the Lord God. Recall that the Hebrew letters for Adam (the second *a* is not a "letter") are ADM, aleph = 1, dalet = 4, and mem = 40, so that ADM is $1 + 4 + 40 = 45$. This harmonic number contains no powers of 2 and so is the harmonic root located in the harmonic matrix as the square of prime number $3 = 9$, which means "third brick from the left," times prime 5, which means "on row two" of any holy mountain. Thus placed, Adam is to be seen as note D of a set of possible octaves, whose tritone will always be the cornerstone, bottom left in his mountain. A unique set of harmonics can develop around such a harmonic root, as a direct function of its location within the matrix, and Adam's root of 45 is a very special one of these locations.

Stories of gods and super-gods often document key harmonic roots and their associated mountains, building a pattern of ideas corresponding to landscapes and astronomical time. The Bible seems to have taken a human-centered tack, focusing on the smallest yet most significant harmonic root possible to propose a story of human development and to explore the potentials of human beings and their desirable limits, in harmonic terms. However, the harmonic root of Adam was also the root of the planetary matrix.

CREATION IN SEVEN DAYS

As with the Sumerian creation story, the writers of the Bible (ca. 600 BCE) needed a prelude to express the intended core meaning for a text relating to the harmonic planetary matrix but without the planetary gods. The chosen form for the prelude appears derived from the Sumerian cosmogony and was a creation—taking place over seven days—of the world of harmony as it operates with prime numbers transposed into waters, dry land, sky, creatures, fish in the sea, and finally the creation of man in God's image. This first story is called Elohist because the name used for these gods was the Elohim (plural), confusingly translated as God in the King James Bible. This plurality of God is symptomatic of harmony with regard to primes, intervals,

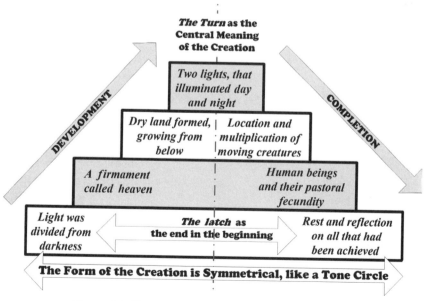

Figure 3.1. The seven days of creation seen as a pedestal narrative structure often used in analysis of biblical stories.

limits, and locations in the underlying structure of just intonation. The story created the framework for the implicit biblical theme of divinity as harmony, in preparation for stories in which harmonic parallels with human ideals and experience would enrich a tribal history. In a theological sense, an eternity (or heaven) was proposed that *coincides* with an existence (or Earth) within which harmonic superposition is then a religious good.

The structure is sevenfold and can be summarized as having a compositional form called a pedestal.[1] Quoting from Genesis:

> In the beginning God created the heaven and the earth. And the earth was without form, and void; and darkness was upon the face of the deep. And the Spirit of God moved upon the face of the waters. And God said, Let there be light: and there was light. And God saw the light, that it was good; and God divided the light from the darkness. And God called the light Day, and the darkness he called Night. And the evening and the morning were the first day. (Gen. 1:1–5)

What the Vedic harmonists might call a *bindu* (the creative point), the Sumerians called Apsu, the "first one":

> When the skies were not yet named nor earth below pronounced by name, Apsu, the first one, their begetter and maker Tiamat, who bore them all, had mixed their waters together, but had not formed pastures, nor discovered reed beds; when yet no gods were manifest, nor names pronounced, nor destinies decreed, then gods were born within them.[2]

When God says "Let there be light," there are lights in the sky, and these have synodic periods that are harmonically related. In Genesis, the light (*aor*) is manifestation within the darkness of the unmanifest, which cannot be known except through phenomena. Just as Apsu is the cornerstone for a creative harmonic matrix, so the Elohim move upon the waters (mentioned in both texts). Genesis then deviates from the

Sumerian creation of gods and instead develops the topography for a harmonic Earth and those who live within it.

> And God said, Let there be a firmament in the midst of the waters, and let it divide the waters from the waters. And God made the firmament, and divided the waters which were under the firmament from the waters which were above the firmament: and it was so. And God called the firmament Heaven. And the evening and the morning were the second day. (Gen. 1:6–8)

This firmament, *raqia* in Hebrew and meaning "expanse," has factors $4 \times 5 \times 19 = 380$, which can also be read as $3 \times 8 \times 10$, giving a close formula for Plato's "4:3 mated with 5," (see chapter 4) equaling 240 rather than Anu's 60. If this firmament divided the waters, made up of powers of 3 and 5, then integer harmonic numbers had been separated from fractions. The generalized firmament for composing a harmonic field employs the powers of two numbers, 3 and 5, as two dimensions within which products define locations within that space. The factor 19 seals this creation as that of YHWH, though specified by the Elohim declaring this framework.

> And God said, Let the waters under the heaven be gathered together unto one place, and let the dry land appear: and it was so. And God called the dry land Earth; and the gathering together of the waters called he Seas: and God saw that it was good. And God said, Let the earth bring forth grass, the herb yielding seed, and the fruit tree yielding fruit after his kind, whose seed is in itself, upon the earth: and it was so. And the earth brought forth grass, and herb yielding seed after his kind, and the tree yielding fruit, whose seed was in itself, after his kind: and God saw that it was good. And the evening and the morning were the third day. (Gen. 1:9–13)

Using the power of 2, the Elohim commanded the powerful waters below this firmament to gather up and form dry land. Such holy mountains, within the firmament, are accompanied by seas, the nonsymmetrical tone areas within a holy mountain. YHWH has three of these, one at each corner, the cornerstone region being the Garden of Eden (see next section). Symmetrical tone sets only occur around a limiting number D, enabling D to reach such tones from both low and high D (the limit). Visually, mountains can be turned upside down and overlapped to show tones at equal vectors from D but in opposite directions, and these are symmetrical and "on dry land" (and called by Plato twins or paired male warriors, wrestling). The seas of nonsymmetrical tones form the rest of a holy mountain. Only the symmetrical tones can form proper modal scales since these form bidirectional component intervals with each other and each ascending mode is another's descending mode, since the modal scales are born twins (see chapter 6).

> And God said, Let there be lights in the firmament of the heaven to divide the day from the night; and let them be for signs, and for seasons, and for days, and years: And let them be for lights in the firmament of the heaven to give light upon the earth: and it was so. And God made two great lights; the greater light to rule the day, and the lesser light to rule the night: he made the stars also. And God set them in the firmament of the heaven to give light upon the earth, and to rule over the day and over the night, and to divide the light from the darkness: and God saw that it was good. And the evening and the morning were the fourth day. (Gen. 1:14–19)

All of the planets, to a lesser or greater degree, are harmonious with the moon and the stars, and so the firmament is hosting these relationships. Since only synodic relationships are involved, the sun is mediating through the Earth's orbit, while the lunar year is a synodic period relating directly to a number of different musical intervals—as with the bridge of a stringed musical instrument. In creating the firmament prior to the celestial objects, the writers are true to the human order of creation in which just intonation had to be algorithmically expressed by man before it can be populated with the celestial objects that conform to its pattern of harmony. Day four is the point of central meaning on the pedestal of figure 3.1, and it is the celestial harmony that is the central meaning rather than the gods.

THE GARDEN OF EDEN

After the seven days, there was life on the Earth but no man to till the ground. Having created life instead of gods, a mist goes over the Earth and a singular Lord God creates Adam out of the clay of the Earth and breathes life into him. He planted a garden eastward in Eden full of pleasant and nutritious vegetation, and the tree of life was in the center of the garden and also the tree of the knowledge of good and evil. The Lord God told Adam not to eat from the tree of the knowledge of good and evil, for eating it would cause him to die. Immediately, God decides Adam should not be alone and creates Eve as companion.

From the perspective of the Jewish writers, the gods have been omitted and they are replaced by the tree of the knowledge of good and evil, whose fruit, when eaten by Adam, causes ejection from the Garden; I would argue for Eden being a harmonic allegory in which the nonsymmetrical region of YHWH's holy mountain is being equated with the ancient Near East, especially Mesopotamia, where the writers (deported upper-class Jews) are in exile. This is a similar metaphor to that used in the Gilgamesh story about the monster of the Cedar Forest (chapter 7), in which regions of a holy mountain have meaning.

The Lord God provided a serpent to tempt Eve into offering Adam fruit from the tree of the knowledge of good and evil. Eating the fruit will open up godlike knowledge of good and evil and make men like gods, yet by eating they are disobeying the Lord God. By placing Adam among the gods, between the lunar year and Jupiter and within range of the easiest of all tritones, Saturn, who forms the cornerstone, the Lord God (known as El Shaddai until Moses) is placing the Adamic race in the midst of the first three rows of harmonic mountains where practical musical modes can form fully around Adam.

THE FLOOD

There seems no possibility of practical musicality until Adam's female descendants have slept with the gods, leading the Lord God to kill off the resulting giant progeny, that is, large tone numbers, with a flood.

This allows us to locate Adam and Eve when the Garden of Eden is viewed in the mountain of new super-god YHWH, with a gematria of $10 + 5 + 6 + 5 = 26$, identified by Ernest McClain as $10^5 \times 6^5 = 777,600,000$. That is, YHWH (as 10, 5, 6, 5) is 26 by addition of its number-letter equivalents, but this name also implies a much larger number as 10 *to the power* of 5 times 6 *to the power* of 5, the fifth power of $3 \times 5 = 60$, when interpreted as in an exponent notation.

The cornerstone region of asymmetrical tones in figure 3.2 has the same number of bricks, nineteen, as the base of YHWH's holy mountain. The cornerstone region is that for the limit 1,080, which in the astro-harmonic matrix is the location of Jupiter, the dominant outer planet with respect to the lunar year. Between Jupiter and the lunar year sits Adam, then, as 720. Opposite from the cornerstone lies a large

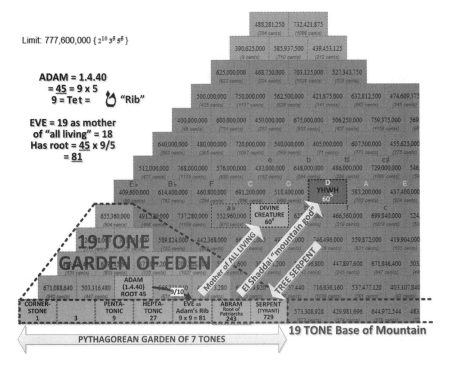

Figure 3.2. The harmonic mountain of YHWH with the Garden of Eden set within the nonsymmetrical region of 19 tones. Only Adam = 45 has a power of 5, the other actors being "on the Earth." The cornerstone is on the left and serpent's tree of knowledge on the right.

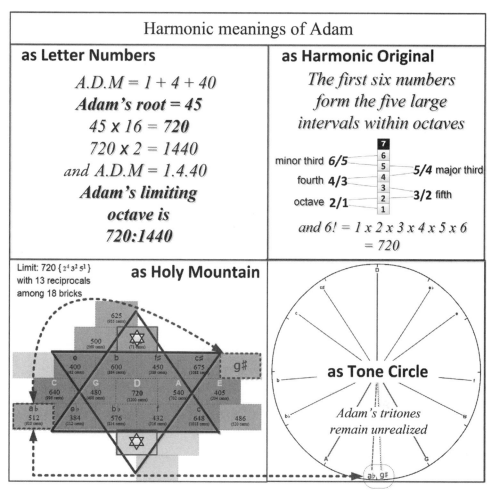

Figure 3.3. Matrix for 720, developing from Adam in the patriarchal line of doubling 45.

region of nonsymmetrical bricks that are loaded with giant powers of 3, namely $3^{10} = 59,049$ to $3^{18} = 387,420,489$, due to the unrestrained cycling of fifths in serpentine fashion.

I suggest that the story of Adam, hinging on his name as a number, now rearranges* the Garden into an ark around Adam as those

---

*The Sumerian and Babylonian flood stories instead sought to destroy the humans created to do the work of the noisy Iggigi, lesser gods working on landscaping Mesopotamia, who supported Tiamat (whom Marduk killed), the leading parts of Tiamat then reused by Marduk to enable just diatonic scales.

limits within his range of 45 to 1,440. If one wants a hulled boat, of a Mediterranean kind, then the limit for 1,440 would give five modal scales, ascending and descending, incorporating a pentatonic central register. The animals come in twos as the chromatic intervals (25/24) between the top and bottom of the boat. The boat can float upward through multiplication by 10 so that, for example, after two more powers of 10, the ark would reach 144,000, the number in the Revelation choir (see chapter 9).

The giants opposite the Garden in YHWH's mountain have their highest (unsymmetrical) brick, at YHWH's tritone G# that, opposite D, represents such upsets (or a change of tonic) like the flood (see chapter 2). If the flood reached that level of YHWH, then the ark could float up to $1,440 \times 10^5 = 144,000,000$, riding the flood caused by G#, then coming to rest on YHWH's mountain. Not eating the fruit, staying within Eden, and the Flood were probably symbols of staying within the bubble of just intonation, as it is far more efficient and better behaved than purely Pythagorean scales. It is the harmonic world that can define a chromatism of twelve tones formed in two ways, but just intonation's triple-layered version is superior over the serpentine, single-layered version. This leaves Noah returning from the heights, as the patriarch of the patriarchs. The latter will take forward the matter of Adam, which would then appear in Canaan, and start with Abram who came from Ur of the Chaldees.

THE PATRIARCHS

Female archetype Eve, "mother of all living," becomes Abram's wife Sarai. Childless, Abram was encouraged (by Sarai) to have first son Ishmael by concubine Hagar, but the Lord God (the mountain god whose number sums to 345) renames Sarai as Sarah and Abram as Abraham. At a miraculous 90 years of age she gives birth to Isaac ("he laughs"), then reportedly dies at 180 years old. It is harmonically relevant that (*a*) the giving of heh = 5 to Sarah and Abraham elevates them from their former selves onto the second row, "stepping up" like the god Ea in Sumeria, and that (*b*) Adam's number 45 has now been doubled to 90 and can form an octaval womb in which Isaac can be born, his life to end at the doubling of Sarah's 90 years to 180 years.

In chapter 1, Isaac's limit of 180 was shown useful, numerically embracing the lunar year and the Jupiter and Saturn synods as 120, 135, and 128, respectively, in units one-tenth of a lunar month. It is also interesting that it is Abram who first follows the god El Shaddai (345), and the gematria of ABRAM sums to $3^5 = 243$, a harmonic root leading to the later revelation (to Moses) of El Shaddai as YHWH, by five powers of prime 5 that then match Abram's five powers of prime 3. It is Isaac's grandson Joseph who came to express the lowest patriarchal limit of 125 required to express the lower heavenly *patrix* (patriarchal matrix) of the synods of Saturn (1), the moon (15), Adam (45), Uranus (125),* and Jupiter (135) with his twelve great grandsons. This implies that the core of the story of the patriarchs was informed by the need to preserve the astro-harmonic matrix.

Jacob, son of Isaac, accelerates rapidly onward with a further doubling to 360 (the schematic year observed in Egypt and Mexico perhaps, but not), and then to 720 "days and nights," again searching for the potential twelfth-note tritone locked up but available as the cornerstone of the matrix. Jacob is renamed Israel by El Shaddai after he has seen a ladder with angels traveling upward and downward. He also wrestled with the angel of the Lord. Jacob/Israel has twelve boy children and one girl, Dinah. As eleventh male child, Joseph is the eleventh harmonic root in ordinal numbers (see figure 3.4) but in the holy mountain for 720, Joseph is 500 matrix units, and the twelfth male, Benjamin, is the location of Jupiter's 540 matrix units in a perfect symbolism of twelveness, present as 135, since Isaac's earlier limit of 180.

Joseph is betrayed by his brothers and sold to slavery in Egypt. His insight enables him to interpret an important Pharaonic dream, foretelling a seven-year famine following on from an equally long glut. He

---

*Third outer planet, Uranus, only appears properly on the mountain for 1,440 where the planet becomes 1,000 matrix units.

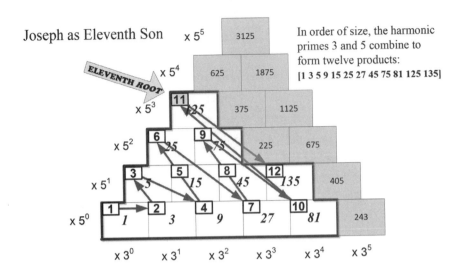

Joseph as Eleventh Son   x 5⁵

ELEVENTH ROOT → x 5⁴

x 5³

x 5²

x 5¹

x 5⁰

In order of size, the harmonic primes 3 and 5 combine to form twelve products:
[1 3 5 9 15 25 27 45 75 81 125 135]

3125

625   1875

11 | 125   375   1125

6 | 25   9 | 75   225   675

3 | 5   5 | 15   8 | 45   12 | 135   405

1 | 1   2 | 3   4 | 9   7 | 27   10 | 81   243

x 3⁰   x 3¹   x 3²   x 3³   x 3⁴   x 3⁵

Limit: 180 { 2² 3² 5¹ }
with 7 reciprocals
among 12 bricks

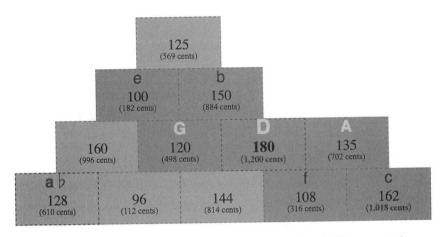

125
(569 cents)

e
100
(182 cents)

b
150
(884 cents)

G
160
(996 cents)

120
(498 cents)

D
180
(1,200 cents)

A
135
(702 cents)

a♭
128
(610 cents)

96
(112 cents)

144
(814 cents)

f
108
(316 cents)

c
162
(1,018 cents)

Figure 3.4. *Above,* the counting of the twelve sons of Jacob. The numerical order in which the mountain's root values arise as products of three and five: 1, 3, 5, 9, 15, 25, 27, 45, 75, 81, 125, 135. *Below,* this outline of "bricks" does exist as the earliest pentatonic mountain, with the limit 144, but is also that of Isaac's age at death, as a limit of 180.

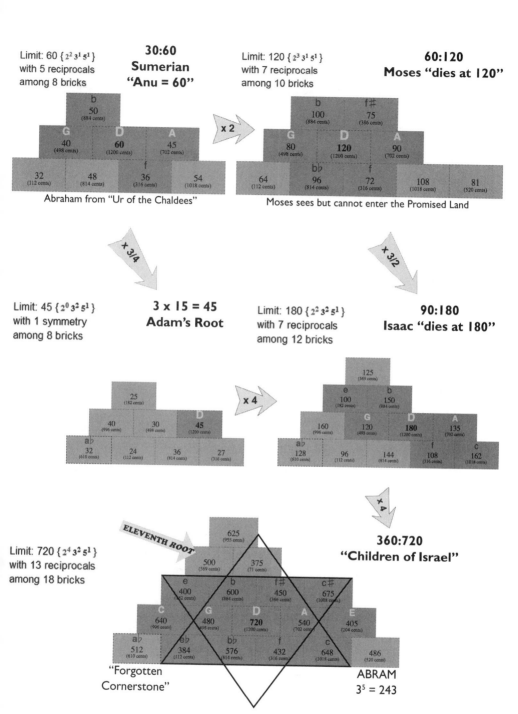

Figure 3.5. The translation of Adam (45 on left) through Isaac (as 180 center) to Israel (as 720) on the right, within which five of the seven modal scales are realized, the star (or "magen" shield) of David excluding C and E, enabling (modern) Phrygian and Ionian (the major scale), each both rising and falling since they are symmetrical (see chapter 6 for scales on mountains).

is appointed vizier in charge of Egypt's grain stores to prevent starvation. This eventually brings his brothers to look for food and meet the vizier and be astounded to see their lost brother, whereupon Jacob's family resettles in Egypt. Joseph's true nature unlocks the patriarchal matrix with Adam's number of 1,440, at which limit Joseph becomes 1,000, entering a new upper heaven where ratios are decimal because he, as planet Uranus, known or unknown, is the origin of the astro-harmonic matrix units that are the Uranus synod divided by 1,000, that is 369.66/1000 = .3697 days; this unit also 1/80 of a lunar orbit.*

Thus Adam's top range, as a limit, is an ideal one for viewing the planetary matrix and can only have been conceived by the writers of the early Bible knowing of the planetary matrix. This then conditions the meaning of their enterprise as both religious and esoteric.

EXODUS

At the end of a period of "captivity" in Egypt, the Bible presents a Moses (777) who will lead 600,000 children, through the positional code 777,600,000, out of Egypt, in its second book, Exodus.† The name of Moses, like that of El Shaddai, sums to 345 and, ignoring other matters, it is to Moses that El Shaddai reveals his true name of YHWH— 777,600,000.‡ It was Ernest McClain[3] who saw YHWH as $10^5 \times 6^5$ = 777,600,000. This number 60^5 = 777,600,000 is one power of 60 greater than that of Isis/Ishtar/Venus as 60^4 = 12,960,000 (Plato's sovereign number), and it seems no coincidence that a god, El Shaddai = 345 (that is 3 × 4 × 5 = 60), should reveal to a man, also numbered 345,

---

*One unit more (81) being 30 rotations of the Earth, located on the mountain at the harmonic root Abram = 243.

†A scribal "translation jest" is involved with Noah's "father," who dies at 777, just five years before the Flood (in Ernest G. McClain, 2002).

‡As already stated, YHWH also sums to 10 + 5 + 6 + 5 = 26, but one has already seen that, in mountain speak, powers of harmonic primes offer data compression and simplicity, since they show you where a number is located on the mountain. Twenty-six can have a double meaning though in the New World, as one-tenth of the *tzolkin* 260-day period (see chapter 8).

that he is $60^5 = 777,600,000$, Moses then leading (as 777) 600,000 children of Israel out of Egypt, since it is the power of 5 that was a harmonic innovation.

The dramatic factors of Moses's birth* led to his adoption by an Egyptian princess and his training in the priestly arts. Egyptian influences were built into the Bible as part of a vital transition toward the new super-god, superseding both Venus (numerically lesser) and Marduk (extravagantly greater) or Osiris (lesser). This seems to present what the Bible's writers needed to say: Abraham came from Ur of the Chaldees (Sumerian harmonic knowledge) and Moses came from Egypt (the pharaonic sciences), remembering again that the stories of Genesis and Exodus were not written down until 600 BCE and that these early books were written in Babylon. It is not that the writers of this great tribal epic betrayed their tribal history through creative writing but rather the rule for writing myths was always to preserve the truth in such a way that it would survive despite being taken literally. The history of the Jews was most likely written by Jewish initiates with all the priestly skills of the ancient Near East, and they sought to present a secret tradition in textual form using allusions to harmony, within both tribal and theological contexts. In doing so, they forged something unique to the ancient world, a dramatic history. The harmonic allusion at the start of the Bible were both human-focused and astronomical, and a conscious shift in emphasis was taking place concerning human development. The religious story would not glorify the planetary system as gods, but would focus on the human creature within YHWH's kingdom. The early Bible appears to have fulfilled two independent aims: the preservation of the planetary knowledge through harmonic allusion and the need for the human imagination to turn away from the gods, and hence toward monotheism.

The wager taken by the writers was of a human evolutionary imperative served better by saving the secret knowledge behind their story,

---

*Moses's mother placed him in an ark-like basket to float off so that a cull of his tribe's babies should not include him. Moses was found among the reeds by the pharaoh's daughter.

a truth that must not be forgotten and so had to be sublimated into a moral tribal history to organize human affairs differently. It is possible to see the Promised Land of the Bible as the domain of sacred human action, likened to the seven modal scales, these "tribes" that require twelve note classes for their total expression.

4

Plato's Dilemma

Jewish harmonic allusion may only have started in Babylon, drawing on their own oral tradition to create a powerful tribal history on which they could superpose an esoteric tradition. The Greeks, in their own preclassical Dark Age, had an oral tradition that, by being written down around the same time as the Bible, gave rise to the literary masterpieces of Homer, the *Iliad* and *Odyssey*, and Hesiod's mythologies and cosmogonies. In their Dark Age, the Greek's mythos seems to have attached itself to the work of the Navigators,[1] a mysterious seafaring people, possibly Bronze Age Minoans or Phoenicians who, from recent analysis of the Greek myths, appear to have plied the eastern Mediterranean around 2500 BCE. At sea, navigation relies on recognizing stars at night, and our present-day constellations were established long before the archaic Greeks and these were then merged with a zodiac largely innovated in Mesopotamia, there developed to synchronize with the seasonal year.

LIGHT OUT OF DARKNESS

By the time of Homer and Hesiod, the myths about the gods were an extensive and coherent corpus, sometimes congruent with other mythologies from elsewhere. This similarity between myths from different regions could have come from some form of cultural diffusion by land or sea. It could also have come from the types of subject we call invariants, such as astronomical or harmonic structural forms, which were then alluded to in unique ways suited to different cultures.

In the case of Greek myths, both harmonic and astronomical allusion appear to be at work and, in Greece in particular, a strong correlation exists between myths and specific constellations and planets, so one might say the Greeks were worshipping the gods through a series of astronomical vignettes rather than through acts of mere human fantasy. In doing so their bards were imagining, composing, and remembering tales in a manner unrivaled today. Education of the young, and entertainment of the people, was by specialists who could recite and perform renditions of the great stories with a lyre accompanying and other emotional and dramatic cues in support. Thus, it is no wonder that a dramatic tradition emerged in Greece and, quite possibly, this is also why reason and philosophy first genuinely emerged there. In its early stages, for example, philosophy was performance, taking forms such as sophism and rhetoric, which led to discussions about democracy, tyranny, or oligarchy.

Perhaps to match the certainties of the myths, repeated and remembered, there were also riddles that came from the gods or the fates, with ambiguous or obscure questions whose answers could actually precipitate good or ill, at the hands of an oracle; for the will of the gods was seen greater than that of man, and to question it was hubris leading to destruction, perhaps after a bout of madness. A key disruption in this self-referential system of thinking was Socrates. Plato on his behalf proposes that the reason of men was being stymied by the performance and audition of myth, filled with unchanging details about what might have been but causing a paralysis in thinking new thoughts about the world and what it was truly made of. It is probably for his trouble that he was executed, but his disciple Plato recorded many dialogues involving Socrates and one focusing on his problem with the Homeric tradition. In a dialogue called *Ion,* Socrates provokes Ion of Ephasos, a notable rhapsode, a performer of Homer from memory, to admit that through mythic performances neither the performer nor the listener is able to reason. While connecting to a hoary tradition, as with iron filings to a magnet, the magnetism of myths stood in the way of original thinking, but Ion has to continue to transmit the tradition he knows; he cannot change.

Therefore, not long after the Bible stories of YHWH, Plato's Ion presented a riddle within this dialogue called *Ion:* "Where is Apollo in Ion?" To explain, Athens was the last eastern city of the Ionians, who had formed the core of Archaic Greek myth, Homer being an Ionian. The answer given by Plato is that "Ion is in Apollo." While literally true, Apollo being an Ionian god, the text of this dialogue is a perfect execution (by Plato) of Homeric meter that is contrived to be 7,776 syllables long in order to be the head number of 60 to the fifth power (777,600,000), the number of YHWH.[2]

In the world of holding numbers in mind, both Jews and Greeks used a loose type of decimal base 10 rather than Sumerian sexagesimal base 60, and the habit was to separate a larger number like YHWH's 777,600,000 into its head number 7,776, as factors of primes 3 and 2, so that a number of zeros we have after the number 777,600,000 (= 5) could be kept in mind as five 10s, possibly using counters equal to 10.*

Plato appears to make Apollo, as the son of Zeus-Jupiter, exactly what Zeus does not want Apollo to be, his successor. In Hesiod's *Cosmogony* we learn that Zeus's father, Kronos, suppressed Zeus and his siblings; likewise Zeus had been warned that a son of his was likely to depose him. Kronos had tried to swallow his children, and Zeus's earth mother had presented Kronos a stone to eat instead of Zeus. Now Zeus, who had loosed two birds from the ends of the world to find its center, had located it in Delphi, which was Apollo's preeminent oracular center. Zeus had himself slain Typhon and buried her beneath Mount Etna, and now Apollo, like Marduk, had slain the serpent Python, whose cult place Delphi had been, thus subverting the native oracle there to Zeus's Olympians.

The country became scattered with *kouroi*, images of Apollo made according to a later Egyptian numerical canon. In the context

---

*Recall YHWH = $10^5 \times 6^5$ = 777,600,000, and in this example the pairing of 2 × 3 = 6 and 2 × 5 = 10 is perfect for highly decimalized large numbers like 12,960,000 (1,296) and 8,640,000,000 (864). In Sumeria YHWH would be big 5 or 60^5 = 777,600,000 and hence simple in a different way. One should realize that exponents or powers of numbers were not a recent invention but existed in Sumeria, since place notation is a form of exponent employing powers of 10 or 60.

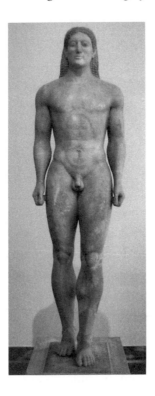

Figure 4.1. Kouros, circa 530 BCE. One of many freestanding ancient Greek sculptures that first appeared in the Archaic period in Greece and represent nude male youths. There were also female equivalents called Kore.

of Marduk's flood that revealed a superior musical scale, employing prime number 5, Apollo claimed to have invented the lyre or seven-stringed scales with which music could better be made. But Plato tells us through 7,776 syllables of *Ion* that Apollo's number is not that of the flood but rather that advocated by the Bible as overseen by YHWH's more explicit code that includes the necessary five powers of 10 and hence of 60.

Regarding the influence of Egypt, remember Moses. Did he not first know El Shaddai as YHWH? Did he not "escape" from Egypt with details of how to build metrological temples such as the Tabernacle and later Temple of Solomon? Hebrew, like Greek, adopted a phonetic alphabet innovated by the Phoenicians, who were the first Canaanites but became the great traders of the eastern Mediterranean, trading to Greece's Aegean Sea. If the Egyptians were influencing Greek statuary and ceramics, and considering that Hesiod's tale of Zeus was as old as the Hittites (who had mysteriously departed just before the Greek

Archaic period)—was YHWH a principle of change sweeping through a hidden world of harmonic thought, turning away from the stories of the gods as such and toward the growth of human reason and doubt, under the banner of Apollo, who respects no established order, including that of his father, Zeus?

It seems Apollo/YHWH was replacing Zeus and the purely mythological form of thinking that had buried planetary harmonic metaphors within such myths, using numerical tuning theory. Mythological thinking was being replaced by something relating to how harmony works within the human world: reason. In human life it is the healthy mind that achieves its own harmony with the world, for which harmony that life is responsible.

HOMER IS DEAD, LONG LIVE SOCRATES!

Plato tells us all we can know about Socrates's ideas and those of others in his world. Plato becomes the custodian, through his writings, for much of our understanding of his milieu and his report on what was replacing the ancient world. The transition from an oral culture to a literate one was almost complete, in the sense of works being available to a whole class of readers. His student Aristotle, tutor to Alexander the Great, was, through his writings, the teacher of many Europeans and Arabs in the following two thousand years. And the most interesting thing about Plato's own output is his expertise in both numerical tuning theory and in astronomy, with some notion that the two are connected. Since the Greek myths of Homer allude to harmony, Socrates is deconstructing the outer face of myth, its slavish doting on fate, heroes, and gods. It was an apparent disrespect for the tales of the gods that got Socrates killed.

The "problem" expressed in Plato's *Ion* can be used to understand why harmony and astronomy were being dissociated from myth in classical Greece. Much of our culture is communicated orally and by memory, then supplemented with texts and calculations that constitute a literary storage of past work, not wholly relying on memory and recitation. This means we are rather blind to the actualities of oral societies in the ancient world who relied much more on sound and speech events informing the function of memory. Even the existence of cuneiform in Sumeria and Babylonia did not make society at large independent of recitation, of hearing and remembering great stories like those about Gilgamesh.

Ancient stories with underlying harmonic references were not written using the medium of writing, but were composed using only the mind, voice, and memory. Homer's oral tradition emerged in Archaic Greece despite cuneiform having been available in the Mediterranean for two thousand years, and used in correspondence between the great empires of the Egyptians, Hittites, Assyrians, Persians, and so on. This is presented as the Greek Dark Age, in which written language was lost when it obviously still existed in the eastern Mediterranean. But, in the vacuum left behind by the Hittites, Greece became home to new Indo-European tribes from the north who, as in the Indian subcontinent, had their own oral tradition composed by spontaneous poets like (in India) Valmiki, who created the epic Rāmāyana through speech.

It is therefore simpler to assume that the harmonic information held within ancient texts came to be there not to preserve knowledge in the distant future, but to preserve it through recitation. When it was written down, people could read it in the distant future but this was not why it exists. Harmonic allusion and references to astronomical knowledge became the natural fare of oral works because oral works were, before writing, the natural medium within which this information had been developed, by people who communicated facts in ways designed to remember them through speech.

Socrates had grasped that reason, rather than memory, was the natural forte of written texts as a permanent externalization of referents, on which one could then operate independently of experience, exactly as a computer program can act on external and stored data, according to the rules of logic. Plato was giving witness to another Adamic moment of departure from the womb-like Eden where the gods appear to create information. Adam must till the earth and make it his own, while Plato was clearly aware that information about the gods and harmony should not, like the proverbial baby, go down the plug hole as not real. This is,

I believe, why Plato is heroic in having given us all the clues we need about what the oral tradition had been remembering in the first place. In the new medium of writing, reason, and logic, Plato has survived, as has Homer, and, together with others, all was not lost at this inevitable shifting from orality and belief to literacy and reason. But only recently has Ernest McClain awoken the sleeping tradition of harmony within the ancient works that were written down.

This explanation points back further than the ancient world to the megalithic period when a culture of not so primitive astronomers developed their skills to the point of recognizing the harmony of the World Soul. It seems the megalithic must have been oral in the deep sense mentioned above, of developing records of their work within composed acts of language while performing observations of celestial events and counting time as distances, developing a metrology that could hold our numbers as lengths and compare time periods using right triangles. They also had a drawn technical art, developed since the Old Stone Age, that could be overlaid by oral recitation to say what it meant.

Orality is therefore the reason why specialist knowledge underlies mythic "literature," which is why the norms of our own literate culture won't let us digest the advanced astronomical culture of the megalithic era. The objectives of oral composition are easy to comprehend. An oral tradition requires a high degree of data compression in a tightly controlled sentence structure. This is exactly what is found in the Bible and other ancient texts, where specialists now exist in the language structures found at the level of stanzas, syllables, prosody, repetition, ring composition, and other archetypal forms. These are sometimes called codes, which run alongside harmonic numbers. Such compositions required highly developed skills, essential to the oral composer in order to add to the cultural corpus. These properties were noted, and adopted when later works were written down, since the scribes could read them to prevent errors in the hand copying of manuscripts.

We see esotericism within ancient texts because we think they were composed by scribes when the type of content and the mechanisms for such composing did not rely on writing. But the skills of the oral culture were *becoming* esoteric by 600 BCE and Ion could remember but not recompose as bards would do. The actual teachings of Jesus (rather than his hagiography) are of an oral nature. Muhammad was an oralist. The movement toward a truly literate society continued and, like a plant at the end of its season, orality could express new flowers through being written down, in signs and symbols and in the nonverbal vernacular of monuments. Perhaps all of Zeus's children were collectively born through writing to exceed him.

ATHENA CAUSES ZEUS A HEADACHE

After the defeat of the Persians in 478 BCE, the temple to Pallas Athena* lay partly in ruins atop the Acropolis. By the middle of the fifth century, Pericles, Athens's most prominent statesman, commissioned leading sculptor and architect Phidias to build a new temple. The nature of Athena is especially important to understanding the Parthenon we see today: she was a transformed image of a more ancient mother goddess, Metis,† and most naturally a lunar goddess who had been adapted to suit the form of patriarchal societies in Archaic Greece. The myth of her creation is to be found embodied in the design of the Parthenon when it speaks of her being born from the head of Zeus, in a strange kind of virgin birth leading to the name of her temple, the Parthenon, celebrating her parthenogenesis. One of the remarkable features of the temple is that her inner cella, like the layout of the city of Jerusalem and the Station Stone rectangle of Stonehenge, took the form of a 5-by-12 rectangle then associated with the solar-lunar calendar, an artifact dateable to

---

*The virgin-birthed goddess who mythically adopted the Athenian city-state, after whom the city was named.

†*Metis* meant cunning wisdom. Metis was the one who gave Zeus a potion to cause Kronos to vomit out Zeus's siblings. She became pregnant by Zeus with a girl-child and Zeus knew her next male child would exceed him and so he swallowed Metis whole, but later Zeus developed a raging headache that Hephaistos relieved by splitting the skull of Zeus and out sprang Athena with a shout and fully armed.

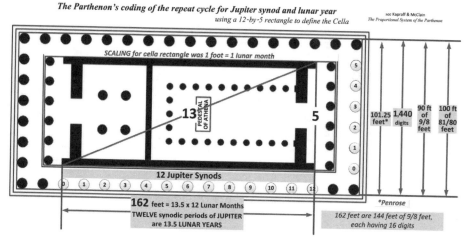

Figure 4.2. The Parthenon's cella, coded as a Pythagorean triangle with sides 5–12–13 and therefore presenting the lunar years of 12 and 13 lunar months plus an intermediate hypotenuse (at 3:2 point) of the solar year, in their relative lengths.

the megalithic period. In addition, the temple's stylobate platform is of such a width, running exactly north-south, so as to define a single second of latitude.[3]

The oldest form of this 5 × 12 × 13 "Pythagorean" triangle (doubled) is found at Le Manio, Carnac, in the west of France and from about 4000 BCE, where day-inch counting had accumulated lengths for three lunar and three solar years (as discussed in chapter 1). The same technology, applied also to the synodic periods of Jupiter and Saturn, reveals that Athena (as the lunar year) relates to these two outer gas giants of the solar system in musical ratios of whole tone (9:8) and semitone (16:15), a fact *unambiguously* manifested in the layout of Athena's cella (figure 4.3). These three symbolisms, of the solar-lunar triangle, the geodetic division of the meridian, and the harmonic design modeling Athena-moon, Zeus-Jupiter and Kronos-Saturn are the embodiment of Metis as the intelligence of the megalithic astronomers and surveyors, assimilated, developed, and reused to define a new type of cosmically relevant sacred space or temenos.

One further surprising link to harmonic theory emerges from the fact that one second of latitude is 1/3,600 × 1/360 of the double

The Parthenon's presentation of the Harmonic Matrix for 1,440

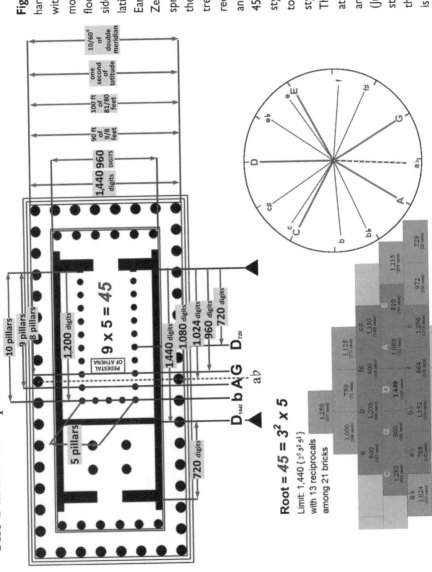

Figure 4.3. The geodetic and harmonic design elements found within the Parthenon. The monument stands on a raised floor (the stylobate) whose short side is a second of a degree of latitude on the meridian, with Earth probably representing Zeus's head, out of which Athena sprang. The longer side hosts the cella and beyond that, the treasury. The cella has an inner rectangle of pillars nine long and five wide signaling the limit 45 while the metrology of the stylobate finds the cella length to be as long as the width of the stylobate, 90 × 16 = 1,440 digits. The eighth pillar is then located at 960 digits (the lunar year) and the ninth is at 1,080 digits (Jupiter synod). Athena's statue stood before the eighth, where the eclipse year lies when 1,440 is doubled, atop the mountain.

meridian around the Earth and that 1/10 of this creates a length 10.125 feet long, which is $1/60^4$ part "of the Earth," and which is the Platonic sovereign number in the *Timaeus.* These geodetic and harmonic relations can then be presented in a further interpretation of the monument.

In typical megalithic style, the measures used create a numerate home for the whole-tone and semitone relationships of the lunar year to Jupiter and Saturn synods that in parallel fit the geodetic division of the meridian. A foot of 9:8 feet was used, 90 of which feet equaling the required width of the monument of 101.25 English feet, which length also equals 100 feet of 81:80—numerically the syntonic comma* that links the Pythagorean and just worlds of tuning, while 9:8 feet, called a *pygme,* is numerically the whole tone between Jupiter and the lunar year. The meaning of the Parthenon as a "hundred footer" (*hekatompodos*) is most clearly related to how 100 feet might form one second of latitude.† But the same distance in digits, 1/16 of the pygme, unlocks the harmonic framework of the cella, since the monument's width and the front two-thirds of Athena's chamber (naos) are both 1,440 digits long.

This 1,440 is the second of the, by now familiar, numerical meanings given to Adam in previous chapters (based on letter-number equivalence where ADM equal to 1 + 4 + 40 is both 45 by addition and

---

*We can see the syntonic comma in the tone circle of figure 4.3, between C and c and between e and E, separating just (lowercase) tones from their Pythagorean (capitalized) tones.

†The geodetic meaning of 100 feet of 81:80 lies in the fact that, in megalithic geodesy, the meridian's length was taken as equaling half the circumference of the mean Earth, the Earth as a perfect sphere and considered the spiritual Earth. Here the Greek geographical foot of 1.01376 feet that would divide the mean Earth in the required way has been reduced to 1.0125 (81/80) feet, which then gives the actual length of the meridian—probably only measured by the Egyptians, who had developed the necessary mathematics to estimate it using a system of algebraic approximation, reconstructed from (undefined) manuscript sources by Livio Stecchini. See my "The Geodetic and Musicological Significance of the Parthenon's Shorter Side Length, as Hekatompedon or 'Hundred-Footer,'" in *Music and Deep Memory: Speculations in Mathematics, Tuning, and Tradition; In memoriam Ernest G. McClain,* ed. Richard Dumbrill and Bryan Carr (London: ICONEA, 2017).

1,440 by decimal placement). Athena is surrounded by nine smaller columns on either side and behind by a row of five small columns, 5 × 9 being 45. There are eight columns up to the back of Athena's own pediment so that the ninth pillar represents Zeus, leaving a whole tone between the two. The holy mountain for 1,440 illustrates where Saturn must be—as cornerstone 1,024—while D = 720:1,440 is the other geometric mean between the lunar year and Jupiter. To review the previous discussion, this makes D the octave between nine and eighteen lunar months, and the lunar year of 960 reveals the framework unit of time, 1/80 of the lunar month—since 12 × 80 = 960. The unit is 0.369 solar days, and this allows the exact astronomical relationship of these Greek gods to be referenced as their geocentric time periods. Furthermore, one finds that 81 such units equals 30 complete rotations of the Earth, showing that the syntonic comma of 81/80 is expressed cosmically between the lunar month and this sidereal month of Earth rotations.

Saturn is placed exactly in between Athena and Zeus, as a root-2 approximation to the "clashing" tritone, which stands "opposite" D, and this point is epitomized as an enigmatic wisdom figure and sometimes fire god Agni, who can see outside the octave framework D presents. Athena's naos or cella is functioning like a musical instrument in which the inner entrance is the beginning of a "string," the pillars are like "frets," and the end wall's far edge is the end of the string. Saturn is soon found to be central in all three dimensions of the Parthenon.

The holy mountain, shown below the plan in figure 4.3, represents the same two-dimensional firmament presented using references to harmonic numbers in the Bible. The Parthenon is centered on the same root of Adam as 45, repeatedly doubled as in the patriarchal story of Genesis, to 1,440. Both monumental Parthenon and biblical Genesis are directing us to the same harmonic pattern that lies in the time periods between Jupiter, Saturn, and the solar-lunar phenomenon we call the lunar year. The designers of the Parthenon could, it would seem, have written the Bible and, with the twin languages of ancient Greek and Hebrew, both of which share the Phoenician letter-number alephbet or alpha-bet, we see the two ventures were similarly engaged with

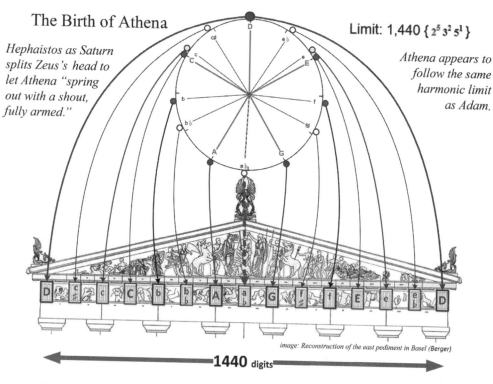

The Birth of Athena

*Hephaistos as Saturn
splits Zeus's head to
let Athena "spring
out with a shout,
fully armed."*

Limit: 1,440 $\{2^5\,3^2\,5^1\}$

*Athena appears to
follow the same
harmonic limit
as Adam.*

image: *Reconstruction of the east pediment in Basel (Berger)*

◄━━1440 digits**━━►**

Figure 4.4. The width of the stylobyte gave an allusion in its length to 1,440 digits as a limit, a length also employed in the cella. The frieze of the eastern pediment is divided into fourteen spaces for sculpture, allowing transposition of this limit's fourteen symmetrical tones, and one asymmetrical a♭ in the middle to express the centrality of Hephaistos/Saturn, splitting the head of Zeus to let Athena out. Note also the seven lintels (called ashlars) bridging the eight columns to then represent the seven intervals of five tones and two semitones: C and E are preferred over c and e so as to achieve two just semitones of 16:15.

the patriarchal social structures emerging in the eastern Mediterranean and ancient Near East, post 600 BCE.

The aforementioned centrality of Saturn within the Parthenon's longest dimension is seen in its harmonic location in between Athena and Zeus, which is the middle ninth outer column (of seventeen) down each of its sides. The second dimension is that enacted by Hephaistos who, like Saturn, was titled Smith, and this is the monument's *one second of latitude* in 100 feet of 81:80 feet—that is, the *width of Hephaistos's axe* used to open the head of Zeus, whose head is now the earth out of which Athena literally emerges "with a shout, fully armed" as the Parthenon.

The third dimension of the monument is expressed through the pediment above Athena's entrance, whose primary Doric frieze is made up of fifteen triglyphs and fourteen metopes (or sculpted tableaux). If we add up the number of symmetrical tones in the matrix for 1,440, there are fourteen (if D is counted twice since D is *tonally* symmetrical and opposite Saturn on the tone circle). When the geometric mean tone of Saturn is added, the total of fifteen allows the triglyphs to present these as a linear octave, as seen in figure 4.4.

THE PYTHAGOREAN PLATO

It seems quite clear that powerful statements regarding the interplay of astronomical knowledge and tuning theory—what I will call *harmonism*—were made in living memory of Plato's summarizing of the Archaic Greek tradition and his consolidation of the emergent tradition, of applying reason to understanding the world. Apart from being a social commentator on the classicizing of Athens and a profound thinker, Plato sought to reconcile reason with musical tuning theory (in the style of Pythagoras) and so came to include a great deal of musical theory in his dialogues, though in a cryptic form that would only come to be properly understood in the second half of the twentieth century. Ernest G. McClain became the focus for this understanding, and he puts it like this in his book *The Pythagorean Plato:*

> Either unknown to Platonists or ignored by them, a truly musical effort to expound Plato's mathematical meaning has been under way for a century. In 1868–76 Albert von Thimus had already anticipated brilliantly the musical implications of Taylor's principle of "perfect symmetry" and Brumbaugh's principle of "cyclic reciprocity" applied to the *Timaeus* creation myth. Von Thimus' ideas were expounded by Hans Kayser and Ernst Levy, and developed more fully in Levy's treatises on harmony. . . . It was my own good fortune

to have been a colleague of Ernst Levy for several years and to have inherited from his fund of priceless insights into Pythagoreanism the particular insight that the formula for Socrates' seemingly impenetrable "sovereign" political number defines what musicians know as Just tuning—idealized, but impracticable on account of its endless complexity—and that the political disaster Socrates foretold must therefore be related to the difficulty musicians have always known. The often-expressed conviction of Hugo Kauder, our mutual friend, that Pythagorean tuning functioned as a kind of "temperament" for the Greeks, was a prescience which has proved helpful in developing Levy's intuition.[4]

This idea of temperament is explored in Levarie and Levy's *Musical Morphology: A Discourse and a Dictionary.* They say:

> The principle and necessity of temperament, properly understood, . . . [involve] forces . . . of growth and limitation. . . .
>
> Two facts are naturally given: the identity of the octave and the otherness of all other intervals. They are irreconcilable except through the compromise of temperament, which permits the formation of closed systems. Octave repetitions set a clear framework but produce no new tones. Projections [of it] upward or downward create "cycles of barrenness" [Plato's phrase in *Republic* viii, 546]. [In contrast,] repetitions of any other interval, however, in either direction continue to create new tones. These are "cycles of bearing." For the creation of a tone system, barrenness is as useless as the other extreme of infinite generation. Temperament reconciles the two by adjusting whatever characteristic interval to the octave. The unmanageable cycle of growth by the power of any interval becomes artificially but necessarily limited by the power of the octave.[5]

When McClain proposed (above) that "Pythagorean tuning functioned as a kind of 'temperament' for the Greeks," he is therefore saying that the Greeks filled octaves with fifths but then tempered them with thirds, to create a just-tuning system. As we have seen, there is a singular pattern that emerged from numerical tuning theory and that can be found in the astro-harmonic matrix, of which Plato, according to McClain, posed the following thesis or theory:

> For restriction to *one* model octave: "Pattern is the Living Being that is forever existent" (*Timaeus* 37c, d). "It was therefore, for the sake of a pattern, that we were seeking both for what justice by itself is like, and for the perfectly just man" (*Republic* 472c). "A city could never be happy otherwise than by . . . (imitating) the divine pattern" (*Republic* 500e). God, the maker of patterns, "whether because he so willed or because some compulsion was laid upon him not to make more than one . . . created one only" (*Republic* 597c). For musical theory, all octaves have the *same* pattern. For Aristotle, the octave was the prime example of a Platonic *form*.

McClain further notes that "Aristotle insists that musical forms . . . are not numbers at all but *ratios;* 'the ratio 2:1 and number in general are causes of the octave' (*Metaphysics* 1013a)."[6]

By suggesting that God, the maker of patterns, "created one only," given the unity of the octave and of temperament, implies that God (however conceived) created the just-intonation system of tuning into which planets appear temporally organized when seen from Earth and measured according to the moon. There is also the thought that harmony inherently leaks into human life with regard to justice and the organization of human affairs, though it may sometimes be that, as in the Bible and other texts, the human world was made a canvas on which to paint harmonic allusions.

POLITICAL ALLEGORIES

Ernest McClain also wrote:[7]

> In political theory as in musical theory, both creation and the limitation of creation pose a central problem. Threatening infinity must

be contained. Conflicting and irreconcilable systems, be they of suns and planets, of *even* octaves (powers of 2) and *odd* fifths (powers of 3), or of divergent political members of a *res publica*—must be coordinated as an alternative to chaos. What the demiourgos has shown to be possible in the heavens, what the musicians have shown to be possible with tones, the philosopher should learn to make possible in the life political. *Limitation*, preferably *self-limitation*, is one of Plato's foremost concerns. His four model cities correspond to four different tuning systems, each with its own set of generators and an explicit population limit:

City	Callipolis	Athens	Atlantis	Magnesia
Character	"celestial" (or "ideal")	"moderate" (or "best")	"luxurious" (or "worst")	"practicable" (or "second best")
Tuning	"tempered"	Pythagorean	Just	Archytas
Generators	$2^p 3^q$	$2^p 3^q$	$2^p 3^q 5^r$	$2^p 3^q 5^r 7^s$
Limit	<1,000	≈20,736	12,960,000	5,040

Plato is saying many things very elegantly, such as that Callipolis, the "ideal," needs only "one thousand defenders," meaning that one can start with the Pythagorean heptatonic of 864 in the Greek Phrygian (modern Dorian) mode (<1,000) to obtain all the white notes of our modern keyboard, a scale that is naturally "twin" to itself in being symmetrical in both rising and falling. If one were to move from D = 864 down to C = 768 as a descending Pythagorean heptatonic (in C major, also <1,000) one needs to invoke the tyrant number as leading semitone 768/729. This is a later and more sophisticated way of identifying the limitations of Pythagorean "temperament" as only having one mode, without invoking the heroism of super-gods like Marduk and Indra to cut and splice the serpent using prime number 5.

Athens, the "best" city, is a Pythagorean enneatonic* system, embracing two black notes, Bb and G#, found on many instruments such as pipes and enabling a limited extension outside the octave range. In his

---

*A city limit has nine-tones when the limit contains four powers of prime 3, enabling four fifths either side of D provided there are enough powers of prime 2, as in this case with 8.

discussion of what Plato has to say about ancient Athens in the *Critias,* McClain defines the

> *Citizens qualified to bear arms.* The "roughly some twenty thou-
> sand" citizens qualified to bear arms in Ancient Athens I read as
> an allusion to the two alternate forms of the *Timaeus* construction.
> "Exponed in one row," the exact limit is 20,736.[8]

With Atlantis (Plato's story still intrigues modern thought), just intonation is proposed as "luxurious" though "worst." This refers to the ancient Near Eastern limit of $60^4 = 12,960,000$, the location of the synod of Venus in the astro-harmonic matrix and Plato's "sovereign number." Like Athens, Atlantis is nine-toned but exalted in just-harmonic space by its fourth power of 5. There are parallels between the Atlanteans who invaded Greece long before Plato and the Persians who had invaded recently: the great empires may have been a target for harmonic ridicule because of their luxuriance.

With the city of Magnesia, the number 7 is allowed to sleep with the other primes, subdividing existing intervals to provide further semitone and quartertone intervals, and these could better approximate important tonal irrationals such as the square and cube roots of two (the octave doubling), hence this city's character is "practicable" and "second best." Its limit is the calendar limit of 720 "days and nights" multiplied by prime 7, and Magnesia was notably studied by Archytas, a good friend of Plato.

MARRIAGE AS AN ALLEGORY OF
SPECIAL NUMBERS

Marriage is the blending of gender components to generate something more than its component parts. It is when the reciprocal implications of scales are confronted that tone numbers, or multiples (of one) require their submultiples (fractions) to be "cleared."* In the result, the smallest

---

*As when one solves an equation with fractions: transform the terms into an equation without fractions—which can be solved—a technique called clearing of fractions.

limit that can achieve a rising scale will need to be multiplied in order to capture, in addition, the descending tone numbers of the same scale in integer form. It is already clear from chapter 3 that the calendar octave based on the harmonic root of 45 is special because, when doubled to form an octave, the cornerstone of all such holy mountains additionally provides 45 with a tritone opposite D, that is 64/32 = 1.4202 of D. As 45 is doubled more tones appear around D and those with counterparts equidistant from D have automatically been "cleared." This allowed the users of such mountains to avoid performing any explicit clearing calculations.

Another special characteristic of octaves based on 45 is their relationship to the first six numbers, 1:2:3:4:5:6, to be remarked on in chapter 5, expressive of all the main intervals found in a modal scale. Multiplied together, these numbers have the product 720, the calendar limit (8 × 45). At this limit five of the crude intimations of modal scales found beneath 81 are automatically "cleared," and become released on the mountain to form just scales. (This significant but relatively hidden aspect within mountains is explored further in chapter 6.)

In the *Republic*, Plato explores justice within the ideal city as "limited to essentials." Pythagoras defined the best tunings as involving only the intervals of fifths and fourths yet just tuning introduces major and minor thirds by employing prime 5, "worst" because 5:4 and 6:5 are incommensurate with 3:2 and 4:3. However, the problems of their incommensurability are offset by the capacity of thirds mixed with fifths to sharpen the Pythagorean semitone (256:243) to the superior just semitone (16:15) and introduce a flatter tone of (10:9), these two being larger and smaller (respectively) by the same syntonic comma of 81:80. This is about maintaining the limitation imposed by the octave that must exactly double. Instead of clearing all the prime 3s with a leimma (lit. "leftover") of 256/243, these two just intervals are clearing the 3s. Also arising is a new capacity, when D = 720, to change mode without changing the tonic, without changing D. It is probably for this reason that the *Republic* asserts: "For a divine birth there is a period comprehended by a perfect number."

McClain comments:

> The perfect number to which Plato alludes, I believe, is the very first one, 6 (= 3 + 2 + 1, i.e., as the *sum* of its proper divisors). The ratios of the first six integers—1:2:3:4:5:6—can define all the tones of the Greek Dorian mode, Plato's "true Hellenic mode" (*Laches* 188d), musical foundation for the *Republic* (398–399), and its *reciprocal,* our modern major mode.[9]

Greek Dorian	D	c	bb	A	G	f	eb	D	(falling)
Reciprocal Dorian	D	e	f#	G	A	b	c#	D	(falling)
Ratios:	1			:				2	
	2		:		3	:		4	
	3	:	5	:	6				
					4	:	5		
		5	:	6					
			4	:	5	:	6		

Table based on Ernest G. McClain, *The Pythagorean Plato,* 20.

The above pattern only becomes available with the limit 720 which is factorial 6,* 6 being a perfect number and hence "limited to essentials." The Greek Dorian is our Phrygian and its reciprocal is our Ionian (being the Greek Lydian). In chapter 6 it becomes obvious that McClain's holy mountain for 720 has five modes available through its symmetrical twins shown as "wetted" bricks.

Any scale possible within the reciprocal tones of a holy mountain is guaranteed through the symmetry of that region to (*a*) exist in both ascending and descending forms relative to D and (*b*) therefore coexist with its reciprocal "twin" scale.

The "necessity" for a modal scale on a holy mountain is that the three vector intervals—T = 8:9, t = 10:9, s = 16:15 (where lowercase again signifies a just interval and *t* means tone while *s* means semitone)—can trace out their pattern as a realizable path within the

---

*Ernst Levy's "tonal index."

reciprocal region. For example, Greek Dorian has the pattern t–T–s–t–T–t–s (falling), and 360:720 offers tone numbers 720–648–576–540–480–432–384–360 that, used backward, give the Greek Lydian, our major scale (rising). The Greek Dorian (rising) is therefore our major scale (falling), and so the two scales are twins, each being the reverse of the other scale's pattern of tones and semitones.

Another set of twins, our Aeolian and Mixolydian scales, can also be seen within the reciprocal area of 360:720, while our Dorian is a twin unto itself, being symmetrical as T–s–t–T–t–s–T. This is the sense in which the perfect number 6 gives, in its factorial, all that is required to a city modeled on it, and I suggest that while marriage is the conjoining of prime 2 and prime 3, as female and male, another type of marriage has occurred between the symmetrical tones, Plato's "ambidextrous fighting males," a marriage between symmetrical modal scales that is only realizable when all of the required reciprocal tone numbers have been cleared, by achieving the lowest possible limit to clear them all.

The modality found in the astro-harmonic matrix is hidden by the fact that the planetary periods are like "sirens" (as Plato asserted), each only giving out a single tone and therefore not conveying music itself, just as a string instrument is merely a set of resonant strings. The significance of just intonation is its ability to *support* modal scales, and hence that must be the significance of the astro-harmonic matrix and the limiting identities of holy mountains in general (a subject discussed further in chapter 6.)

THE HARMONY OF THE SPHERES

All have heard this expression about the heavens, which has become something of a cliché. The phrase was influential for Socrates and Plato, becoming an intellectual framework for prescientific cosmology. Aristoxenus (fl. 335 BCE) referred to his immediate predecessors as "harmonists," a term I use and that means "harmonic theorists." Isobel Henderson comments that *harmonism* has been a divided field ranging "between high mathematical method, the empirical and the inductive."

She continues:

> The term "Pythagorean" is loosely used to cover a long tradition of mixed doctrine. . . . It must be taken at its own valuation, as a self-propelled science, inspired not by a special interest in the musical art but by a general interest in the nature of the universe, seen under the strongly mathematical bias of Greek thought. Its aim was to reach a theoretically satisfying scale, which was conceived as a structural element of the *cosmos.* The astronomical firmament was pictured in the Music of the Spheres, from whose revolutions was emitted a scale of tetrachords, each divided by two 9:8 tones with the *leimma,* or "remnant," of the perfect fourth (Plato, *Timaeus* 35b). The Pythagorean di-tone was really used in classical [i.e., Archaic] music, but not long after it was obsolete the austere scale of the Spheres played on—not to the sensual ear but in manuals which recorded it by sheer force of theoretical tradition. Astronomy remained a regular branch of harmonics. The attempt to express the universe in numbers admitting of an irrational element was not absurd in itself, though it lacked experimental method and finally descended into morasses of Neopythagorean mysticism. It served as a hypothesis to stimulate much first-class mathematical work, which was carried on not only by "Pythagoreans," but also by such scientists as Ptolemy.[10]

The noted lack of either practical application or experimental method was probably because it had lost its origins, as the astronomy of the day was turning to mathematical methods and it was often Babylonian tables that substituted for the astronomical observations and the counting of days with which the megalithic builders had been able to deduce the necessary evidence. Much of the numerical methodology preserved in Plato and the other harmonists, evolved in the ancient Near East, was becoming mathematics, algorithmics, analysis, and theorems. The basis of harmony in factual astronomical terms was lost, and cryptic texts only helped confuse the matter.

The poverty of surviving works on practical music from before the current epoch is attributable to the method of oral transmission, natural within the oral societies that performed, for example, Homeric works. It was also true that Homeric musicality became unfashionable in classical Greece and that some works may have been destroyed, intentionally or not, by later Roman or Christian cultures and certainly by cultural vandalism such as occurred with the destruction of the library at Alexandria. Sometimes words retain an original significance, such as *syntonos*, meaning "taut," implying a tempering of Pythagorean strings, so as to enlarge the semitone to its just interval of 16:15 while then reducing adjacent whole tones within the tetrachord with one 10:9, its just equivalent. This requires a tuning change by the *syntonic comma*, a change of 81/80 accomplished through tightening a string on the harp, as discussed in chapter 5.

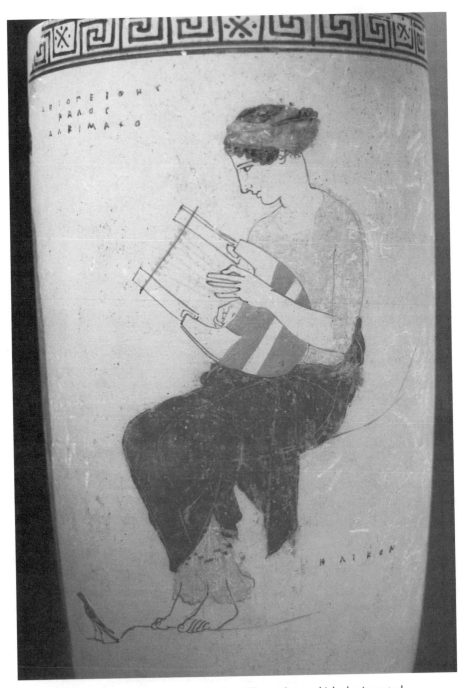

Figure P2.1. Muse playing the lyre. The rock on which she is seated bears the inscription Hēlikon. Attic white-ground lekythos, 440–430 BCE. Painted by the artist called Achilles.

PART 2

∽৩৩৩

A COSMICALLY CREATIVE HARMONY

A stronomy and harmony are both invariants, the first cosmic and the second belonging to the abstract world of number. Invariance, when recognized within a message from the past, can defeat the inevitable losses ancient information suffers due to the passage of time in that an invariance can be reconstructed from a small, fragmentary reference to it. Musical harmony and the relationships between planetary periods are both to be found in the human sensory world and are the same today as they were for those observing them in Sumeria or in the Neolithic. In the case of musical harmony, this invariance flows directly from the growth in size of harmonic numbers, relative to one another, when ordered according to increasing magnitude. In contrast, the planets are parts of a gravitational system that must have *become* harmonious, then conforming to just intonation, the ideal tuning system if pure tones are to be respected. Hence, the musical harmony recently manifested between the planets and our moon forms a pattern of harmonic synods that realizes, as a musical instrument might, the seven modal scales (see chapter 6) rather than some mere concoction of tones and semitones belonging to scales.

Consciously or not, Sumerian tuning manuals (chapter 5) presented how the seven scales are interrelated, "on earth as it is in heaven," to our acoustic music. The harmony of the outer planets would have seemed, in the holy mountains presented earlier, a celestial harmony encouraging life to also become harmonious on the holy mountain of the Earth.

In part 1, Saturn was seen to be as opposite D in the tone circle, and hence a discordant tritone to the implied tonic (home note or key) of an octave's D, its god or "Deity." This is a threat to D's leadership that does not arise when a lyre is first tuned to the modern Dorian scale using pure fifths and fourths, which are easier to discriminate than tones and semitones. The tipping of the Dorian's horizontal tritone, when tuning to other scales, causes the tritone to approach Saturn's vertical position, at which point further tuning can "topple" D itself. This tuning method was lost to history until recently, once texts were deciphered and their invariant subject of tuning identified. The Sumerians discovered that naturally populated octaves have their tone sequences arranged in seven different ways, conveniently organized by the location of the tritone, a location unique to each mode as it rotates around the notional tone circle of the octave (chapter 5).

The Sumerians therefore conceived of the modal system as being created by the tritone that, eventually—that is, after cycling through the modes—opposes the tonic of D, then overthrows D, thus preventing the tonal dominance by any pitch of the governing octave interval. In other cultures too, the tritone came to be seen as a god. In separate visions, the Sumerians saw Ea-Enki as "god of the waters," the Vedic seers (in northern India) saw a sacrificial fire god called Agni, the Greeks saw a Phoenix bird arising from its own ashes while, in the World Soul, it is Saturn between the lunar year and Jupiter. The tritone god is an archetype derived from the invariance of musical harmony. If so, many ancient civilizations have been touched by the source of that archetype, but is this because of musical theory or a cosmically creative harmony established in heaven?

Chapter 6 looks at where the seven scales are to be found on the holy mountains introduced in part 1, developing familiarity with the scales and how scales appear to have influenced north Indian thought as found in the Rg Veda. India developed a still renowned modal music as Europe lost its musical traditions in the first millennium, and Indian cosmological systems reflect their tuning theory as based on just intonation. The rishis of the Rg Veda probably originated the flood-hero myth,

Development of Sri Yantra

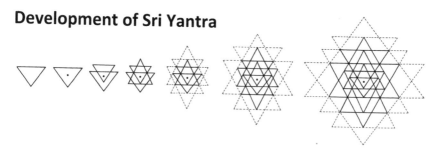

Figure P2.2. The Sri Yantra, a visual tuning text.

constructing large harmonic mountains such as for 432,000 "years" for the Kali Yuga within a Maha Yuga system of 4,320,000 "years" since the universal flood-hero number is 8,640,000,000. The ancient Indian Sa-grama scale involved twenty-two *srutis* from which the seven scales naturally emerge under the different limit for a holy mountain of 4,320 and 8,640, three times larger than 1,440 and 2,880. In fact, the iconic Sri Yantra could have been a visual tuning text for the method of the Sumerians, developed early in the current era as a vision of the goddess Tripura Sundari, underlying the World Soul.

5

The Quest
for Apollo's Lyre

Before setting out to write this book I bought an eight-string lyre, the instrument associated with Apollo, Greek god of harmony. I needed the lyre to understand some tuning instructions probably written by the Sumerians between 3000 and 2500 BCE. Written in the oldest written script (cuneiform) on clay tablets, the instructions are now five thousand years old and have only recently been found and understood to be musical tuning texts.[1] The Sumerians lived in the south of "the Land of Two Rivers," ancient Mesopotamia, now part of southern Iraq. They invented civilization and the technologies associated with city living, leaving behind records on clay tablets, inscribed with a syllable-based script with which they then also innovated written history. Since 1970 at least a dozen "mathematical texts" involving numbers have been interpreted as (*a*) instructions for the tuning of harps like the lyre or (*b*) the study of musical intervals between harmonic numbers. It seems Sumeria had a profound understanding of musical tuning, both by ear and by intellect (that is, through numbers.)

Reconstructing Sumerian and later Babylonian musical knowledge is made more difficult because our own tuning standards have, in principle, abandoned pure musical ratios. To be able to play together, modern instruments are tuned according to a compromise strategy for

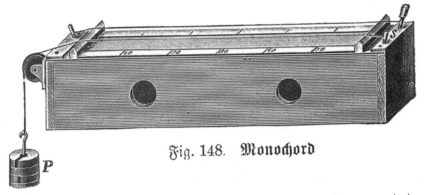

Fig. 148. Monochord

Figure 5.1. In a monochord, one or more strings, fixed at both ends, are stretched over a sound box while one or more movable bridges are manipulated to demonstrate mathematical relationships between sound frequencies. "With its single string, movable bridge and graduated rule, the monochord straddled the gap between notes and numbers, intervals and ratios, sense-perception and mathematical reason." David Creese, *The Monochord in Ancient Greek Harmonic Science* (Cambridge: Cambridge University Press, 2010), vii. Image: *Bibliothek allgemeinen und praktischen Wissens für Militäranwärter*, vol. 3 (Berlin, Leipzig, Wien, Stuttgart: Deutsches Verlaghaus Bong & Co., 1905).

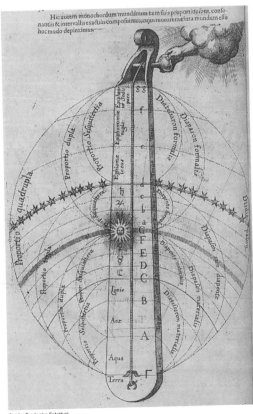

Figure 5.2. "Music, mathematics, and astronomy were inexorably linked in the monochord." Siemen Terpstra, "An Introduction to the Monochord," in *Alexandria: The Journal of the Western Cosmological Traditions*, vol. 2, ed. David Fideler (York Beach, ME: Red Wheel/Weiser, 1993), 138. Image: "The Divine Monochord," illustrated by Robert Fludd (1617), showed three realms: material (sublunary), ethereal (astral), and empyrean (eternal).

tuning called equal temperament. This makes it impossible to appreciate the ancient Sumerian instructions without a lyre or similar instrument that allows strings to be tuned freely so as to be harmonious by ear and hence use pure musical ratios.

That Sumerians had an advanced harmonic tuning system by 2500 BCE contradicts our own foundational myth about music: that it was the Greeks who first developed the tuning systems needed to develop a numerical theory of harmony.

Both Pythagoras, "the father of musical theory," and later classical Greeks inherited their knowledge from the many older empires of the ancient Near East—the Sumerians being the first, the last being the Archaic Greeks, whose "Dark Age" from 1200 to 800 BCE was without a written history but rich in oral myths and legends—notably Hesiod and Homer, whose works were performed by a harpist or rhapsode. The eventual reinvention of civilization in ancient Greece came about from at least three sources: the matrilineal traditions of the Cretans and the Aegean Sea, the incoming Indo-European tribes such as the Ionians and Dorians, and continuing contacts with an ancient Near East and the Phoenicians. The Near East was still rich with civilizing ideas such as musical theory and religious mystery, which is why the myths of Pythagoras take him there to discover those mysteries. What was new about the Greeks in the first millennium BCE is their particularly rich mythology, including Apollo as one of twelve Olympian gods formed around Zeus-Jupiter, who (according to Hesiod, the Dark Age mythologist) had deposed his father Kronos-Saturn.* I believe that in Apollo's myth, the lyre represents a foundational form of harmony inherited, not from Greece through Pythagoras, but from Sumeria and its later derivatives, based on their study of the internal structure of the octave doubling (or halving) of vibrations.†

---

*Note that now Jupiter is a whole tone replacing Saturn's semitone relative to the lunar year, which was the basis for the Dark Age calendar.

†And Iamblichus says Pythagoras brought tuning knowledge back from Babylonia.

THE MYSTERIES OF THE OCTAVE

The eight strings of the lyre can be tuned so as to populate a tone circle where a single rotation around its outside circumference represents a doubling of the rate of vibration between strings one and eight, the number eight giving us our word *octave* and hence implying the eight-string lyre. Octave doubling is a numerical relationship between two vibrations in the ratio of 2 to 1, and this relationship is the primordial *interval* in harmonic tuning theory. With our intellect we can prove, as Pythagoras did, that the two smallest numbers, 1 and 2, are involved in creating this interval of the octave by exactly halving the length of a string on a monochord. The relation between string length and rate of vibration is reciprocal: the

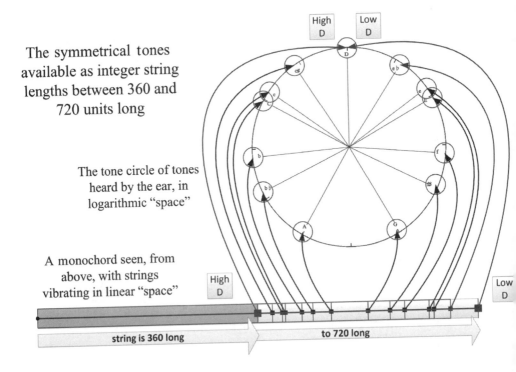

The symmetrical tones available as integer string lengths between 360 and 720 units long

The tone circle of tones heard by the ear, in logarithmic "space"

A monochord seen, from above, with strings vibrating in linear "space"

string is 360 long   to 720 long

Figure 5.3. The logarithmic tone circle and equivalent monochord view. The backbone of this tuning system is Pythagorean (uppercase letters) deriving from prime 3, while the new just tones (lowercase) are caused by prime 5, when tuning theory can accommodate these within limits such as 720. Notes C and E both have alternative just tones, employed by different modal scales.

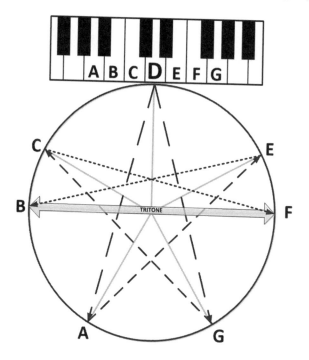

Figure 5.4.
Logarithmic tone circle for the generation of the Pythagorean heptachord, created as a starting point for tuning a harp to just intonation's seven scales (see figure 5.6, page 105).

higher tone produced by halving a string length causes it to vibrate at twice the frequency of the tone of the whole string.

Inside a tone-doubling circle, other harmonious intervals can be located by ear that then define how the eight strings of a lyre should be tuned. These intervals can similarly be found as ratios on a monochord or by using numerical tuning theory, for these intervals involve ratios where odd numbers greater than 2 are splitting the octave, by division, into eight parts. Such intervals, ordered in continuously increasing pitches between 1 and 8, can be made to form seven possible orientations of the same pattern, always involving only two sizes of interval between adjacent strings, called tones and semitones. These emerge naturally by finding (ascending and descending) fifths and fourths (figure 5.4), starting from (by definition) D, and resulting in the "base" tuning, called the Pythagorean heptatonic, or in modern music called the Dorian scale, symmetrical about D (see preface to part 1). The Sumerian scribes recorded a fundamental process for retuning this Pythagorean diatonic scale of seven notes into a series of seven different-sounding diatonic scales that improved the Pythagorean semitone (in the system called just intonation). Just tuning sounded better and was more versatile, creating the modal music of India, for example.

TABLE 5.1

The intervals in the Dorian scales found in Pythagorean and just-tuning systems.* Note the displacement of two 9/8 intervals by 10/9 and two 256/243 semitones by 16/15.

Intervals	T	S	T	T	T	S	T
Pythagorean	9/8	256/243	9/8	9/8	9/8	256/243	9/8
Just	9/8	16/15	10/9	9/8	10/9	16/15	9/8
Intervals	T	s	t	T	t	s	T

*Because of its cyclic generation from D, the Pythagorean heptachord is forced into the single orientation that obeys the inherent symmetry of the tone circle (see figures 5.4 and 5.6 for how a harp is tuned to it).

Tuning the lyre strings to the Pythagorean scale, the ear is very prone to *incidentally* correcting the semitones to being the more harmonious 16/15 interval, their just-tuned equivalents, sharpening 256/243 by 81/80* to 16/15. But one would not then come to the orderly sequence of seven modal scales that came to dominate Near Eastern and modern music. These seven forms (called modes by the Greeks and church modes or scales today) arrange their five tones and two semitones by flattening one of the tones by the same amount that its adjacent semitones is sharpened—a vital step that ensures the octave remains an exact doubling or halving in pitch.

Players resolved sequences of notes separated by five whole- and two half-tone intervals within melody because the structure of the octave naturally prefers this structure when tuning with fifths. This tuning of multiple strings led the Sumerians (or their antecedents) to an advanced understanding of how number affects the form of modal scales, *counterintuitively*, through the location of the worst sounding interval, called the tritone. While harmonic experiments on string length do reveal the numerical foundations of harmony, such experiments never

*This difference, which relates these two tuning systems, is the *syntonic comma.*

gave the West what the Sumerians gave the ancient Near East: a vision of how the harmonic *scales*, as well as harmonic intervals, emerge naturally from the domain of small numbers, through the location of the two opposing semitones. It appears that the change in human thinking by the time of Greece was not sympathetic to the contemplation on the pure forms revealed by Sumerian tuning procedures.

The Greek myth of Apollo represented the lost harmonic knowledge of the Sumerians, while the myth of Pythagoras (who disdained just tuning) represents the rational approach to harmony at the close of the Greek Archaic period. As the "light" of written history returned to Greece, the works of Homer and Hesiod, products and perpetuators of an oral cultural norm, were written down to become literature. Harp playing accompanied recitation and learning in upper-class schools, giant sculptures of Apollo complemented the myths and temples, and exquisite geometrical decoration of vases marked a period of high cultural achievement. But this oral tradition was slowly turning into a culture driven by and expanding through literature through the new Semitic alphabets and in a few centuries the classical Greece of Plato and Aristotle was the result, making Pythagoras the very end of ancient music rather than its beginning.

The common misperception is that the ancients found music an essential subject of study alongside geometry, astronomy, philosophy, and so on, but this is a medieval notion invented through the few texts extant in the West after our own Dark Ages, these amplified by the romanticism and egalitarianism of the last two centuries. Music has also been transformed by its now long classical tradition, our numeracy and physics are now in the public domain, and our religions are still conflicted with each other and with secular society. These overlays have to be overcome by knowing what one can't assume.

The central quest here is to ask the question: What is music made of and why? While the ear discriminates harmony, this does not reveal how it does so, and so Pythagoras contrived an unmusical study of string lengths, where all other factors such as string tension and thickness and density have been made identical. String lengths then revealed the underlying numerical ratios responsible for harmony. In contrast and much earlier, the Sumerians studied the invariant forms the pattern of eight-note scales could take, when semitones are never less than two tones, or more than three tones, apart.

A lyre tuned using fifths and fourths finally arrives at a discordant tritone because the fifths and fourths arrive at a semitone that is unplanned, called by the Greeks the leimma or "leftover." This semitone was merely left over after cycling through fifths, by ear, to make seven different tones within the octave. The tuner is going to tune to 16/15 instead of the leimma, by stealing 81/80 from the adjacent 9/8 tone, so as to make that tone 10/9 (see figure 5.5 *embubu*). If corrected by ear, F and B will move inward, toward each other, within the octave.*

By now moving F at one end of the resulting tritone, and enlarging the semitone to make a tone to f#, the tritone is moved on to being C to f#, and not eliminated. The end of the new tone combines to make a new tritone, this time to the other end of the opposite semitone (b–C). The Sumerian brilliance at this point was to realize that, in avoiding the tritone, (*a*) a new mode (figure 5.5 *pitu*) has been created relative to D that leads to different-sounding melodies but that (*b*) far from breaking the opposed semitone/tritone situation, two semitones still face each other as before, and so the same trick can be repeated to form another new scale (figure 5.5 *nid qabli*). This repeatability of a similar situation remains true until, as we see later, a maximum of seven different scales become possible—thus inventing *modal diatonic* music.

The strings of the lyre allowed the Sumerians to comprehend the seven modes as the invariant children of diatonic octaves, generated by ear while defined by the number field. The tritone between the two opposed semitones is the creative power that makes this possible. Their approach made the lyre a crucial extension for tuning theory, operating on a more holistic footing than Pythagoras's monochord. It was perhaps

---

*When sounded together they will still make the tritone, which is an interval three whole tones apart; in scale terms, it is the augmented fourth moving from F to B, or the diminished fifth moving from B to F.

no coincidence that Homer's rhapsodes accompanied the myths they intoned with the harp.

In the physical world, things that vibrate *can* relate to each other harmoniously when they are coupled and sympathetically resonant, just as vibrating piano strings cause collateral vibration in neighboring strings. Harmony is therefore a type of systems theory in which vibrations can feed backward and forward between resonators. Systems that can vibrate like strings can take on harmonic intervals toward other

Effect of just tuning by ear on the semitones and adjacent tones

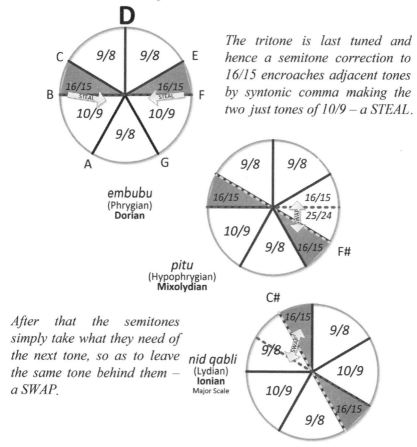

The tritone is last tuned and hence a semitone correction to 16/15 encroaches adjacent tones by syntonic comma making the two just tones of 10/9 – a STEAL.

After that the semitones simply take what they need of the next tone, so as to leave the same tone behind them – a SWAP.

Figure 5.5. Three tone circles showing how lengthening of alternate semitones can create new modes by shifting the tritone implicit in diatonic scales.

strings that then vibrate through resonance rather than through being struck. It is this idea of relatedness within systems that underlies how harmony can operate beyond musical performance, between instruments and voices, between planets.

PRACTICAL TUNING

The simplicity of Sumerian scale shifting through single acts of tuning reveals a deeper and older tradition than Pythagoras's monochord, the Sumerian method not evident in Greek music except as the myth of Apollo's lyre. I noticed in attempting to tune the lyre that one arrives at a satisfactory octave with five tones and two semitones. Like Pythagoras I was searching for harmonious tones by ear but, instead of varying string length, I was varying string tension.* But I could not measure string lengths, as Pythagoras could, to establish the numerical interval ratios between the lyre's strings. But I saw that for tuning scales, eight strings were better than one despite the fact that the human ear is not very good at discriminating the tone and semitone intervals within an octave for the purposes of tuning, leading to tuning by fifths and fourths, which are easier to hear.

The tuning ear must use the most harmonious intervals available between strings (the fifths and fourths), which is why tuning manuals existed in the ancient Near East. The lyre player wants the strings ordered in ascending pitch, so the tuner must tune using ascending and descending fifths (3/2) by skipping intermediate strings from low D up to fifth string A and from high D down (a fifth) to fourth string G, and so allowing the central portion of the scale to be established first. Thus we have the two D strings one octave apart, and the higher D tunes down by a fifth to G, and the lower D upward by a fifth to A (as with the World Soul, where the lunar year is G and Jupiter, A).

---

*In practice, instrumental strings have a nominal length and are then provided with a tensioning method, to "tune up" to other strings, and instruments. Once differently tensioned, string lengths are not in any clear sense proportionate to pitch, and one cannot therefore quantify the interval ratios sounding between strings according to the length of their sounding parts.

Scale Order:	**D₁**	E	F	G	A	B	C	**D₂**
Tuning Order (from the center of D)		F	C	G<**D₂**	**D₁**>	A	E	B

Further tuning is done through treating G and A as new starting points to find C and E, which need to be fourths taken in the opposite sense, so as to remain within the chosen octave. Tuning in ascending and descending fourths from G to C and A to E establishes whole tones at the top and bottom of the scale. Then, from C, a descending fifth will make F, and from E, an ascending fifth will make B. When complete, one has a lyre tuned in the modern Dorian mode, the Sumerian *embubu,* within which B and F form a tritone amenable to giving full access to seven modes via the Sumerian method of tuning.

One can see that the above process could continue to fill all twelve note classes, numbered in figure 5.6 as possible strings of a twelve-string lyre in which all of the notes in the octave, including the black keys of a piano, would be present. But, such a rush into manifesting new tones would not be rewarding, as is also the case with the monochord, where Pythagorean tones increasingly become inharmonious to one another until, at the twelfth note, two different notes, G# and A♭ (which are separated by the Pythagorean comma), appear, whereas these should be the same note to the ear.

By choosing eight strings for the lyre, its human creators had understood the need for five tones and two semitones, seven tones in all, within an octave. This could have been grasped using a monochord: in order to make a practical instrument, eight strings are required rather than the monochord's one, and the possibility of sounding two tones simultaneously enables the tritone to sound, and this opens the door to the creation of the seven modal scales.

In the lower part of figure 5.6, the Pythagorean heptatonic scale is in the Dorian mode, whose combination of tones and semitones is T–S–T–T–T–S–T, or in notes on the eight-string lyre, D–E–F–G–A–B–C–D. The tritone is the interval between F and B. As noted above, the continuous moving of the location of the tritone enables the systematic retuning of a lyre into any one of seven modes, as shown in figure 5.7. Looking at the row marked embubu/Dorian, the tritone F–B has a semitone at either end of it, that is, the intervals E–F and B–C. By tuning one of semitones at the ends of the tritone up to a whole tone, a new mode is created, and the position of the tritone shifts among the notes of the scale to a new place, creating a new mode. In the Dorian mode, if we tune the F up* to an F# (figure 5.7, row 3 to 4), one is making the interval from E to F# (a whole tone) and the interval from F# to G becomes a semitone, and we end up with the D Mixolydian mode, whose notes are D–E–F#–G–A–B–C–D and whose scale steps are T–T–S–T–T–S–T, with the tritone now between the F# and C. The retuned string has redefined a new tritone, in relation to the other string of the opposing semitone. The tritone has been shifted to achieve a new mode, but the tritone always retains the opposing semitones at both ends (as per figure 5.5) and so can continue to perform this procedure until a cycle of seven modes returns again to the Dorian, but then on a tonic one semitone different, higher or lower as to which end of the tritone is retuned, this full cycle being shown in the tone circles of figure 5.10 (page 113).

Something new is also born with the lyre's initial tuning. The method of tuning by fifths yields the Pythagorean heptatonic scale in

Figure 5.6. (*opposite above*) Tuning a lyre to the Dorian mode. All twelve possible note classes are shown, and the tuning starts at high and low D tuned to the octave (*top*). Descending and ascending fifths, respectively, strike 5 (G) and 7 (A). To stay within the octave, fourths in the opposite sense strike 10 (C) and 2 (E), and fifths from these achieve 3 (F) and 9 (B), which are a tritone apart from each other. (*opposite below*) Harmonic matrix and tone circle for limiting number 864, produces the Dorian mode natural to the Pythagorean tuning of seven notes (or heptatonic). The limit appears in the bottom row because it has no powers of prime 5. The darkened bricks indicate tones that are symmetrical within the tone circle. The number 864 is the lowest number able to resolve the Dorian mode.

---

*It is very important to note that in recent scholarship, habits can lead modal scales to be viewed as centered on a tonic of c, while here I hold to D and the symmetrical Dorian scale as the rational place to start tuning, especially considering the doctrine of holy mountains, symmetrical tone circles, and the significance of the tritone opposite D in McClain's vision. The word in Sumerian for "tightening" has also been thought of as "loosening," thus mirror-imaging the ordering of Sumerian scales. We are not involved in such scholarship but reveal how the tuning system of the Sumerians was relevant to our story whether tightening or loosening was meant.

Tuning a Lyre to Pythagorean Dorian Mode

From which Sumerian scale shifting can access all seven modal scales

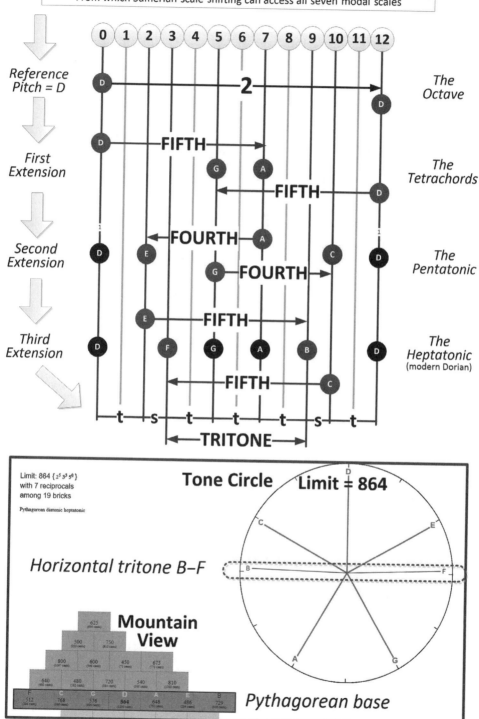

Sumerian Scale Shifting about the Tritone

Revealing the organization of modal scales, seen within numerical tuning theory

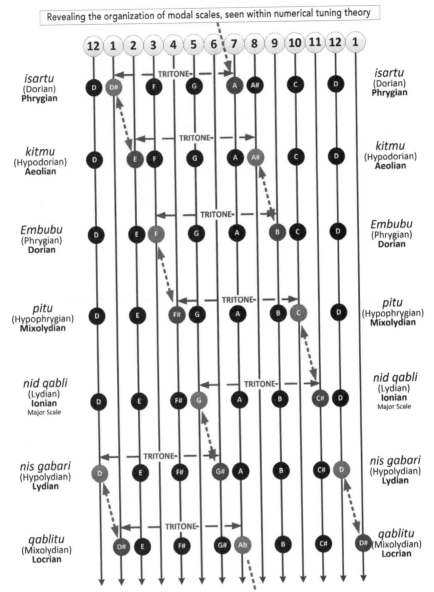

An 8-string lyre uses 7 of the 12 note classes, D twice in forming the octave.

Figure 5.7. The eight-string lyre uses only 7 of the 12 modern note classes (1 to 12) at one time to make a given scale. This figure shows all the possible note classes and the Sumerian method of retuning just one string by a single semitone (gray arrows), in order to move to a different scale. The scale names in the margin are sumerian on top, ancient Greek in middle, and modern below. The thirteeneth note is equal to the first, allowing D (the tonic) to move between modern Lydian and Locrian. The equivalent tone circle view of this is available in figure 5.10, page 113.

which the semitones are called leimmas, and these "leftovers" are *each* a ratio made up of eight powers of 2 (256) divided by five powers of 3 (243), purely because that is what is left over from the octave doubling after all the five whole tones have been generated using successive fifths and fourths. This ratio is approximately 20/19, but when the harp tuner's ear seeks a better sounding semitone, the leimmas will gravitate to the most harmonious semitone interval, which will be 16/15. At that point, the lyre has already entered just intonation, and this movement, using the ear, corresponds to the movement by intellect of blending the prime number 5 with the Pythagorean prime 3, to create modal scales. There are therefore three stages to the achievement of just tuning.

> **First,** the monochord allows for the study of harmonious intervals with regard to their ratio by employing a single string that can have constant properties. A movable bridge varies the vibrating length, and the lengths can then be measured. A five-tone structure emerges with two semitones, inevitably in our modern Dorian mode, involving Pythagorean intervals, and these don't involve prime 5 in its differential string-length ratios.
>
> **Second,** the lyre offers an ability to play notes independently on its eight strings and achieve the same Pythagorean Dorian scale. The tritone (B–F) is a discordant interval joining the two semitones; both sit adjacent to the tritone but outside its range. Tuning one end of the tritone to make it consonant with its former partner achieves both a purer semitone of 16/15 (the ear preferring smaller numerical ratios) as well as a new mode, and this principle is then found to be repeatable through a cycle of seven modes.
>
> **Third,** it would have then been possible to understand the tones of the new tuning intellectually, and comparison with a monochord would reveal the new semitone as being 16/15 and, in order for the semitone to have grown, the tone next to it must have shrunk by the same amount (which is 81/80) to a new tone interval of 10/9. This rationalization comes with the monochord's ability to be programmed with numerical intervals, using the harmonic number sets graphed in figure 5.8.

One of the first of these retunings of the lyre from Dorian to Mixolydian, from F to F#, illustrates how the act of retuning reveals a competing interval to the fifth (that is, of 3/2 and its reciprocal fourth of 4/3) called the third of 5/4, three of which do not quite span the octave (just as seven tones of 9/8 would exceed the octave). The 5/4 ratio in partnership with the Pythagorean 3/2 ratio can obtain a sweeter semitone and enables the seven modes to come into existence on the lyre. It is this system of tuning that is called just intonation, only roughly defined as the combined effect of fifths and thirds. The lyre is naturally disposed to produce that type of tuning, using the Sumerian tuning technique. The tones and semitones are the consequences of successive fifths and thirds, their children so to speak, which caused the Sumerians to deify the properties of 3, 4, and 5 in their product 60, equaling Anu, out of which creation came, a creation based on musical harmony (see chapter 2).

NUMERICAL ORIGINS OF THE SEVEN MODES

In the ancient Near East, mathematical studies[2] were made of numbers having prime factors 2, 3, and 5 (*harmonic* or "regular" numbers). The Sumerians had innovated a number system with a sexagesimal, or base-60, structure rather than our own decimal base of 10. Using a base with all three harmonic primes allowed these primes to be more visible within numbers written using (in loose-place notation) sixty signs. I will continue to use the decimal base to show what would have been seen in sexagesimal.

Looking at the harmonic numbers between 1 and 81, one sees a remarkable natural expression of the seven modes indicating the fundamental structures of musical harmony based on the number field. The Sumerians would have been able to see their scales in terms of octaves below a minimal limiting number, different for each scale. The Ionian (their *nid qabli*) could be realized on a string length with a limiting number as little as 48 units long, while the Locrian (*qablitu*) would need a string length as long as 90 units.

It is then possible to see how these modes might all appear in the same octave, by noticing the requirement that each mode's highest number be a factor of a larger common limit or multiple.

The lowest common multiple of the primitive modal scales, 48 to 90 in figure 5.8, limits any ability to manifest the seven modal scales. The limit 2,880 is the least limit able to provide all of the primes found as factors of those seven limits—see table 5.2.

TABLE 5.2

Modern Mode Name		Earliest Limit Factors
Locrian	90	$2 \times 3^2 \times 5$
Aeolian	80	$2^4 \times 5$
Mixolydian	72	$2^3 \times 3^2$
Lydian	64	2^6
Phrygian	60	$2^2 \times 3 \times 5$
Ionian	48	$2^4 \times 3$

Lowest common multiple = $2^6 \times 3^2 \times 5 = 2,880$

Five of these scales appear together under the limit four times (two octaves) less than 2,880, in the "calendar year" limit of 720 days and nights, referred to by Plato and in the Rg Veda as connecting the world of musical harmony with the organization of time. There is not a large difference between 360 days and 365 days, and in fact the difference is the interval ratio of the Pythagorean comma that separates G# and A♭: an audible mismatch of what should be a single twelfth tone opposite D. It is only when octave 360:720 is doubled twice to 1,440:2,880 that all seven modal scales come into play, and just intonation can reduce the unfeasible "giantism" of the numbers implicit in pure Pythagorean tuning down to a limit of 2,880. The last two modal scales, Locrian and Lydian, are then realized in which g# and a♭ are available, similarly separated but never played together. See chapter 6 for more on scales, but in "religious terms," in subsequent chapters, this exposure of the tritone opposite D within a scale may be equated to the forbidden fruit and the crack in the world egg, when seen as the tone circle.

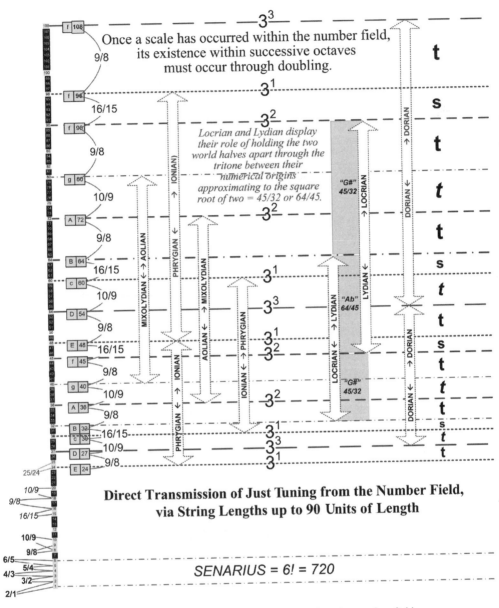

Figure 5.8. The appearance of modal scales within the number field below 81. On the *left* are numbers increasing as a pile of black squares with light harmonic numbers punctuating them. These harmonic numbers define the tones and semitones familiar to diatonic scales. The scales are inflexibly ordered for realizing music but rather prefigure the scales appearing in usable form on harmonic mountains as regions within limits that clear these numbers as multiples and submultiples (fractions) as per table 5.2.

Template of Modal Music
prior to the "creation" of scales on mountains

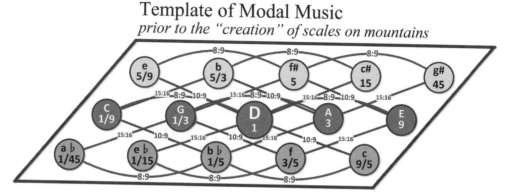

Figure 5.9. The state space for just intonation within the harmonic matrix, taking the limit for high D to be equal to I, shows all of the interval trajectories needed by the seven modal scales. The space required is $3^2 = 9 \times 5^1 = 5 = 45$ but doubled six times, $2^6 = 64$, so as to realize the Lydian mode (see chapter 6).

The planetary system is modeled on this same numerical archetype, where Saturn provides a♭ and g# is provided by the synodic period of Chiron, the largest current asteroid of the Centaur class found between the orbits of Saturn and Uranus.

THE CRACK IN THE WORLD EGG

We have seen that in three expansions, of tuning by fifths and fourths, the tonic (D) lies at the top of the tone circle, between the two world halves (figure 5.4). The lyre arrives, from two routes each made up of two fifths and one fourth, to either end of two diminished semitones called leimmas or "leftovers" between E and F and between B and C (figure 5.5). It is these semitones, created last in forming the Pythagorean heptatonic, that lie opposite each other, linked by the near tritone of B–F. The tritone is at right angles to the axis of symmetry for the tone circle, between the "world halves," where G#/A♭ would act as tritone, opposing the tonic D; this is a circumstance only found in Pythagorean tuning after all twelve notes are articulated through six expansions by fifths and fourths. While the Pythagorean tritone (of the World Soul) cannot form without six expansions, the third expansion generates two tones opposite each other (C and F), at right angles to the World Soul's tritone. This property of the Pythagorean tritone gives us access to a mode of tuning that can:

1. Simplify the leimma ratios of 256/243 by having them steal 81/80 from two of the whole tones, so making just tones of 10/9 and making them more harmonious.
2. Develop the seven different modes through rotation of the tritone until it matches the vertical axis of symmetry of the Pythagorean tritone.
3. Evolve the tonic of D by raising or lowering its pitch by a semitone, perhaps explaining the dictum that the way up is the way down.

Since harmony projects the organization of the number field on a continuum of possible pitches, it only eliminates disorder by providing seven that, in theory, provide a discrete spectrum of harmonic tones and scales, all anchored on a chosen reference pitch. Even in equal temperament, where intervals are not integer ratios, the pitch A (below middle C) is equal to 440 Hz and is an anchor for global music making.

Where pure intervals are sought, combining the factors of three different prime numbers, those primes are forever *incommensurate* to one another. The prime numbers three and five cannot divide into each other and yet they are more efficient and flexible than Pythagorean tuning in populating an octave to form the rich harmonic framework of just intonation. This possibility, of using a limiting number containing the prime number 5, arises naturally through the tritone formed by a Pythagorean, seven-note octave, a diatonic prototype of the five tones and two semitones of which the modal scales are made and a perfect starting point for generating the seven modal scales of just intonation through "clearing" its tritone and later, the tritone present in each modal scale.

The tritones, linking two opposed semitones, provide a simple way for these semitones to be displaced relative to the tonic of D, so as to

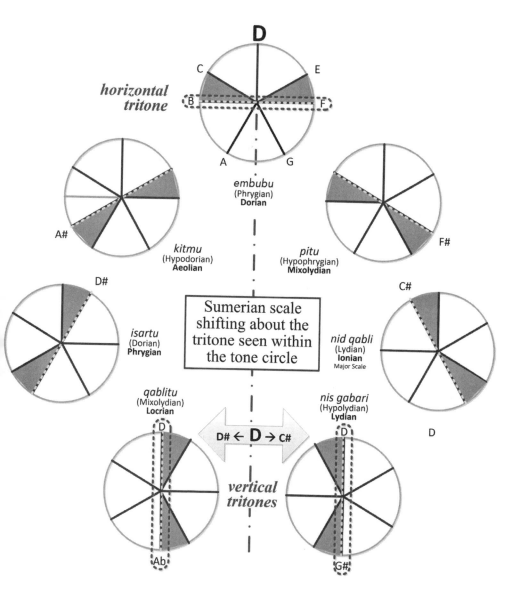

Figure 5.10. The symmetry revealed within the Sumerian scale-shifting process based on the rotation of the tritone and adjacent semitones. The lyre is first tuned to embubu (*top*) from a defined D. The near tritone between B and F (*dotted*) has two semitones on either end (*grayed*). To "clear" the tritone one can tune F to F# (pitu) or B to A# (kitmu) and a new pair of semitones appear in each case at the end of the new near tritone. Repeating the process, *nid qabli* or *isartu* and then nis gabari or qablitu bring the tritone to the vertical position, where further tuning will topple D to D# or C#.

form seven operational modes through tuning a single string at a time, and when the tritone reaches the location of D–G#/A♭, then retuning D can cause both semitones to flip across the world halves, between the Lydian and Locrian modes. This also increases or decreases the tonic by one semitone and is the crack in the tone circle's world egg. This "crack" in the world egg is found in many traditions including Vedic India, Greece, Egypt, and China.[3]

PHYSICS OF THE WORLD SOUL

In the lyre, each string vibrates as a *simplex* (technically a 1-simplex or line or *degree of freedom*), between the nut and bridge, against which the string is tensioned. The lyre has a number of strings vibrating in this way, forming a polychord, rather than the monochord's string(s) of a single note, and its many sounding strings allow the lyre to be compared with the many differing yet harmonious "vibrations" between the moon and the other planetary bodies.

However, in the case of an outer planet its vibrations (or synodic periodicity with respect to the Earth) are already duplex (a 2-simplex or area) in that planetary orbits (their sidereal periods) are subtly altered by the Earth's own orbital period (our solar year). This extra complexity is then further modified by the moon's orbit of the Earth being fully illuminated by the sun just twelve times a year. The harmonic relationships of outer planets to our moon are therefore a triplex phenomenon (a 3-simplex or volume), based on

    (*a*) their own orbital (sidereal) periods,
    (*b*) the Earth's own orbital period (solar year), and
    (*c*) the synod of the moon and sun through illumination seen from the Earth (the lunar month and year).

The harmonies of the World Soul are therefore not one-dimensional like a string, but emerge out of these three dimensions, forming a volume where factors *a*, *b*, and *c* result in a harmonic interval between the lunar year and a planet.

The physics experienced on the Earth cannot provide us with a similar system of harmony to this one found in the sky. We build harmony into our musical instruments, while it takes objects as large as the sun, planets, and moon, widely separated in their orbits that (within a vacuum) are essentially friction free, to achieve a situation in which the regular interactions of planets with each other through attraction and proximity can be the only arbiters of the interval ratios found between them. Neither the simplex sidereal periods nor their duplex synodic periods display any arrival at musical harmony with respect to the Earth. It was only when the moon was created and its orbit lengthened that its *triplex* contribution came into harmony with three outer planets seen from Earth; this harmony is based on the lunar year of twelve illuminations, the same number as the note classes that naturally populate the world of the octave.

It is not unreasonable to compare the triple expansion of tuning by fifths, leading to the heptatonic tritone with this triplex nature, to the World Soul. Indeed, Apollo had both a lyre and a cubic altar, and the altar was associated with a geometer's challenge: to double the volume of the altar's cube by increasing the length of its sides. (Note first that this doubling is similar to that of the octave, but then seen as in volume.) The challenge has no integer solution, but in the ancient world an accurate formulaic approximation using integers was developed: to increase by 5/4, the major fifth, then a further 125th part, that is, by 126/125. This procedure to our eyes is multiplication by 1.26 and it gives a highly accurate doubling of the altar's volume, to better than one part in 5,000.

In India, being able to stride across the world in three strides was an ability of Vishnu, and Apollo's problem viewed musically can be seen solved on our keyboards, where an equal-tempered major third spans four keys, one-third of the twelve in an octave. On the tone circle this is seen in the symbol of the equilateral triangle set within a circle, as in figure 5.11.

In ancient tuning theory, three thirds of 5/4 fail to reach the octave by a dissonance of 125/128, called the minor diesis, yet this ratio is very important for the astronomy of the lunar year's relationship to the inner

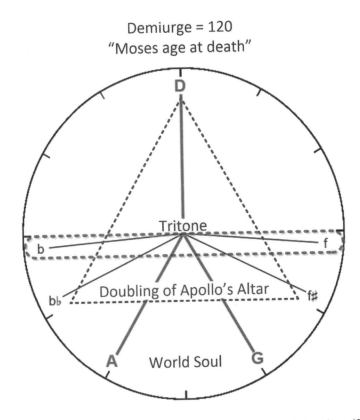

Figure 5.11. Triangle within a tone circle World Soul expanded to limit 120, the age of Moses at death, illustrating the tritone B–F, and the three expansions of the cubic altar by three equal-tempered thirds, f# and b♭.

planets that are raised up above the moon by a minor diesis. It is ironic therefore that in solving Apollo's riddle concerning his altar, equal temperament leaves us incapable of forming the minor diesis because its major third has been enlarged to become the cube root of octave doubling, by Apollo.

Apollo stands, like the Earth's relationship to the solar year, between the lunar year and the synods of the nearest outer planets, including Zeus-Jupiter and Kronos-Saturn, with the justly tuned lyre as vehicle of the tritone's creativity that interconnects modes within a seamless pattern. Apollo's cubic altar represents the triplex nature of the World Soul. And these are only the most recent Greek garments for a mystery found in India, the ancient Near East, and Mexico, where similar heroes pointed to similar truths but in their own inimitable fashion.

6
Life on the Mountain

Why should harmonic mountains have been important for religious storytelling, even if they conform to the system of planets within which we live? The answer may lie in seeing where musical scales reside within harmonic mountains. Our note classes speak of twelve possible landing places for the tones within octaves, near to each other within instruments to allow tone and semitone movements to higher or lower pitches, forming melodies or chordal harmonies.

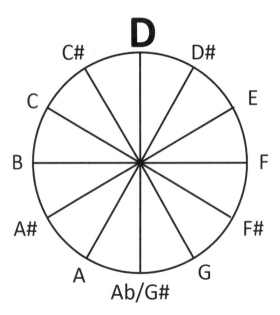

Figure 6.1. The twelve note classes.

SYMMETRIES BY EAR AND BY NUMBER

A particular modal scale is a subset of seven of the twelve note classes, consisting of five tones and two semitones. Other modes are then possible, but in practical modes the semitone interval pairs always stand opposite one another on the tone circle, forming a tritone between them.

The modes are seven basic sequences of tones and semitones, in either ascending or descending order. But as the Sumerians observed, these modal scales exist as symmetrical pairs (or twins) depending on whether the tritone is "cleared" by raising or lowering one or the other end of the tritone. The ascending version of one of these pairs uses the notes of the other twin's descending scale (see figure 6.3) and vice versa.

This results in a pattern among the modes in which there are three pairs of reciprocal scales and a seventh scale, the modern Dorian, that is *symmetrical to itself;* the Dorian being the basis for our note-letter system of D–E–F–G–A–B–C–D or as intervals T–S–T–T–T–S–T— and there are no sharps or flats. In Dorian, the semitones (S) are symmetrical in their placement, and hence the scale is the same whether

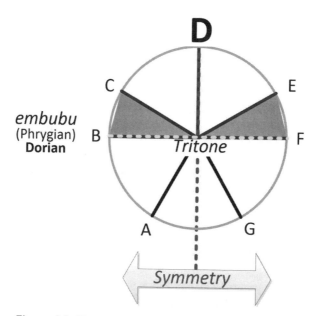

Figure 6.2. The symmetrical Dorian scale, Sumerian embubu.

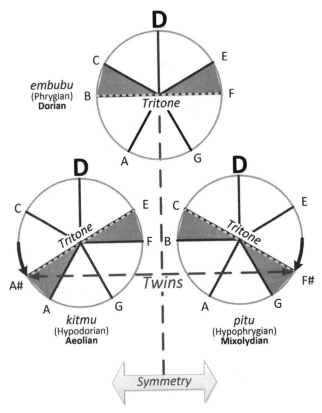

Figure 6.3. The symmetry of twin scales.
The Sumerian and equivalent Greek (*in parentheses*)
and modern (*bold*) scale names are given.

ascending or descending or, in other words, Dorian is its own "twin."*

The Dorian scale and its note letters (with no sharps or flats), present "God on the Mountain" as D = Deity—a limiting number whose opponent lies opposite on the tone circle as A♭/G#, between G and A. But the tritone opponent to the tonality of D is not so easily generated since the Dorian has its tritone horizontal. As discussed in chapter 2, the ancient super-gods could rise up the mountain to improve the tritone made by the cycle of fifths, generating twelve (chromatic) notes with

---

*It is worth mentioning again, the Dorian scale with D as the root was adopted by Ernest McClain for presenting tone circles because it is the original Pythagorean heptatonic and is a naturally symmetrical scale.

small limiting numbers and hence a tritone between D and G# and A♭ that two of the seven scales require—namely, the Lydian and Locrian scales. Marduk and Indra use seven powers of 5 to kill the python and then "justify" the restacking of the serpent vertically, forming the just diatonic. The biblical harmonists covertly presented the ability of limits based on Adam (= 45) to form the cornerstone as tritone when D is placed on the third brick (= 3 × 3) of the first row (= 5) within just intonation. D = 45 is on a vector within the mountain that enables this tritone to be reached from D so as to form the twin scales of Lydian and Locrian. In this one can see that the prime-number factors can be used as vectors to move around the world of harmonic mountains that each have, as limiting numbers, a god = D = Deity on the mountain.

As shown in figure 6.4, one could consider the cornerstone to be a kind of artillery piece in our hands, which takes composite powers of 3 and 5 (a regular number) and throws them, by their prime number content, to the only place they can ever "belong" on the mountain. But then, the harmonic root of that Deity, pumped up by powers of 2, swells into an area of symmetrical tone pairs (twin tones), all these reachable by the same two intervals from the D at both ends of its octave. It is therefore the powers of 2 that allow surrounding numbers within the mountain's octave to be reached from D using symmetrical ascending and descending, or reciprocal, intervals. Once a tone is symmetrically reachable from D, stepping-stone intervals of tones and semitones emerge between it and other symmetrical tone pairs (as if by magic) and these start to connect up tones to form fully functioning and symmetrical *twin scales*. The horizontal tritone of the Dorian mode can then be tilted plus or minus one semitone, creating new twin modes that in ascending provide the descending mode of their twin scale.

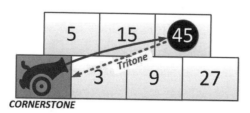

Figure 6.4. Cornerstone projection onto the mountain.

SEVEN SCALES MARRY THE SEVEN TONES

One must follow the biblical harmonists' cornerstone projection of D = 45 (Adam) as the key example, as Adam doubles through the patriarchal series of 90:180:360:720:1,440. By 720 the cornerstone has become 2^9 = 512 so as to fit within 360:720. However, for the cornerstone to become symmetrical, the octave must be able to realize its twin, g# requiring an integer g# = 2,025, and hence two more octave doublings to 2,880 are required. For this reason, 360:720 cannot lead to the creation of those two modes that require a♭ and g#, while it can generate a just form of the Dorian (symmetrical with itself), the pair called Mixolydian and Aeolian, and the pair called Ionian and Phrygian, the other five scales of the seven.

The possibility for full manifestation of all seven modes was perhaps deliberately thwarted by limiting Adam to an upper limit of 1,440. Only the Dorian, the Mixolydian-Aeolian twins, and the Ionian-Phrygian twins can be found in the symmetrical regions of 720 and 1,440 and hence these are Adam's "lot" as far as scales are concerned.

The Sumerian grasp of tuning (chapter 5) embraced the final pair of modes, Lydian and Locrian, *requiring* g# and a♭, and this would imply the Sumerians knew of the limit 2,880, if their numerical understandings were as profound as their tuning insights. And ancient Near Eastern "mathematical texts" do show Akkadian prowess, inherited from the Sumerians, over numerical tunings of an extravagant and demanding sort, handling very large numbers.[1]

Harmonic mountains, because of their exacting and invariant nature, are forensic tools if and when the tuning methods of ancient cultures can be deduced from harmonic number references. Ernest McClain, who pioneered this technique, did not discuss where his system had evolved from—his Pythagorean influences, Ernst Levy and Siegmund Levarie, expanded with his encounter with Vedic specialist Antonio de Nicolás. It was perhaps obvious to McClain how scales worked within his holy mountains, but they were not explicitly mentioned. For instance, what is the real difference between a♭ and g#?

Mixolydian and Aeolian Modes as Symmetrical Twin Scales

These scales employ "Pythagorean" C and E to give the natural pentatonic two diatonic variations.

(*below*) Two tetrachords from D provide a practical framework for positioning the semitone within the pentatonic that does not have one: (*on right*) 2 to 5 o'clock (b–C or bb–A) in the tone circle and (*on left*) 7 to 10 o'clock (f#–G or E–f) positions.

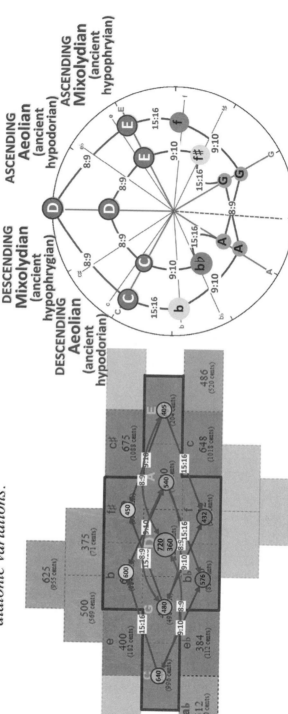

Figure 6.5. Mixolydian and Aeolian modes within the mountain for 720. The variation between these twins is focused on where the two semitones are placed between E–G and A–C.

David's Shield as a Template for Phrygian and Ionian Modes

KEY HARMONIC FACT: 720 is the lowest regular number able to generate symmetrical falling and rising Dorian modes over two overlapping templates of seven brick values, pivoted about D.

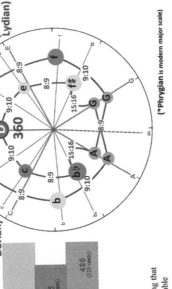

1. To generate tones for both ascending and descending versions of these two symmetrical scales, the mountain is overlapped by an inverted mountain, shown at right as an area of darker symmetrical tone numbers. The ratios between the darker bricks are "ambidextrous"; reachable by high and low D, they allow an ascending and descending scale through the octave, based on brick numbers, and realize symmetrical Phrygian and Ionian scales.

2. These wet bricks can be established visually realizing tones in opposite directions to each other, relative to D, must each be equidistant tonally from both high D (the limit) and low D (limit/2).

3. One can check that both ascending and descending scale tones remain integers by performing a calculation, as below, that is by the method of "sub-multiples."

4. It is then possible to resolve a scale order of bricks, recognizing that just diatonic tones (9/8 & 10/9) and a semitone (16/15) are available throughout such a mountain, between bricks separated by the same vector displacement with respect to each other, as below.

5. The same scale sequences can then be shown (ascending from D = 360 and descending from D = 720 = limit) within the logarithmic tone circle view.

(*Phrygian is modern major scale)

Richard Heath © 2015-2016
Based on the work of Ernest G. McClain
See also http://HarmonicExplorer.org

Pivoting Calculation

CENTS	brick	divide into 360	times 720	CENTS
0	360	1	720	1200
112	384	15/16	675	1088
316	432	5/6	600	884
498	480	3/4	540	702
702	540	2/3	480	498
814	576	5/8	450	386
1018	648	5/9	400	182
1200	720	1/2	360	0

Figure 6.6. Phrygian and Ionian modes within the mountain for 720. The semitones are either the fifth and twelfth position (Phrygian or major scale) or the first and seventh position (Ionian).

"Bed of Ishtar" as Template for Rising and Falling Lydian/Locrian Modes

limit = 2,880 is the lowest regular number able to generate symmetrical falling and rising modes incorporating the 12th tone, opposite D.

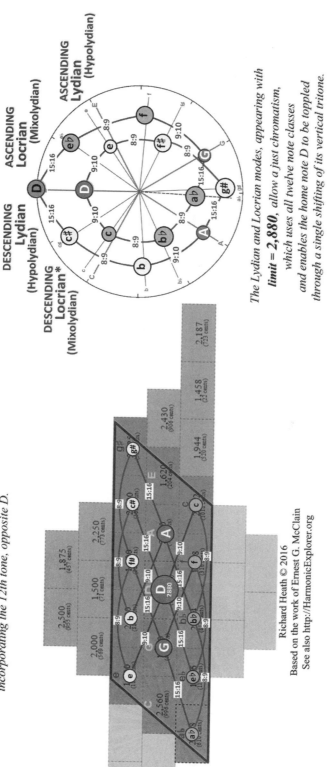

The Lydian and Locrian modes, appearing with limit = 2,880, allow a just chromatism, which uses all twelve note classes and enables the home note D to be toppled through a single shifting of its vertical tritone.

Richard Heath © 2016
Based on the work of Ernest G. McClain
See also http://HarmonicExplorer.org

Figure 6.7. Lydian and Locrian modes within the mountain for 2,880. Mountain (*left*) and tone circle (*right*) both show a♭ and g♯, once symmetrical, permit the sixth and seventh scales to manifest within the mountain as a knot work of intervals connecting symmetrical tones.

This question may seem small-minded, if not ludicrous, within a world where g# is an a♭ by definition! But, in the less-abstract world of ancient tuning theory, there is an important difference: a♭ *rises* 16:15 from G, whereas g# *descends* by 16:15 from A.

If one recalls the World Soul introduced in part 1, note A = Jupiter synod, and note G = the lunar year. The lunar year rises to a♭, the Saturn synod, then tritone to D. In the Lydian mode (which appears alongside the eclipse year of 1,875 in the twin-peaked mountain for 2,880), g# first becomes available as $45^2 = 2,025$, a semitone of 16:15 *below* Jupiter (note A), associated with the Centaur asteroids. This then pairs with a♭ that ascends a semitone of 16:15 above the lunar year (note G). At this point all the seven modes have become symmetrically available. Once these tritones to D have been established, all the seven modes in all twelve keys become available for the "first"* time.

The biblical harmonists, understanding the unique nature of 45 as a vector from the cornerstone, saw the modes coming into existence through the smallest possible limits. This enabled the biblical tradition to be founded on the very profound properties of symmetry available at the cornerstone of a much larger harmonic space, marking this as the ideal location for a human being modeled on just intonation, in "an image and likeness." The cornerstone in fact holds this property around itself as a relic† of the number 1, as seen in figure 6.8, where 1 is actually marking the division between four regions, two pairs of reciprocal regions that interact to manifest scales when the "bed of Ishtar" template is located with D = 1.

Here we have "desiccated" the numbers by stripping any 2s, while showing that when factors of 2 are added, one can see just three types of intervals held within the pattern, two different tones and a semitone. The cornerstone is projecting this pattern into any location on the mountain, being the ghost structure that octave doubling of a projected harmonic root can develop to generate modal scales around D, or any other tone number. This leads to cosmological thoughts like

---

*By "first" one means: the lowest possible limit for g# to appear, which is 2,880.
†Every limiting number, being an integer, divides by 1.

Template of Modal Music
prior to the "creation" of scales on mountains

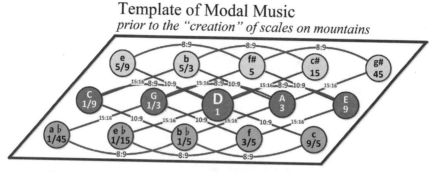

Figure 6.8. Template of Just intonation at the cornerstone [repeated].

those of post-Vedic India, in which creation emanates from the bindu, a creative oneness. And I would propose that numbers greater than 1 are actually 1 (a whole) divided by a number. Seen within a physically oriented mind space, in which lengths are extensive for the purpose of acting on each other, numbers stretch out as a continuum, but in the intensive view every whole is made up of rational parts, like an integer made of ones.

The Octave's Relationship with One

1. Every *whole number* is that number divided by 1, but we see this in reverse in the measured extensive worlds of string length and its reciprocal, pitch.

2. Every extensive *harmonic number* is therefore 1 divided only by the three harmonic primes: 2, 3, and 5.

3. A *harmonic root* is divided by only the factors of primes 3 and 5 in order to locate a possible octave within a two-dimensional matrix of cross multiples, with successive powers of 3 running horizontally and powers of 5 running vertically.

4. A *harmonic development* is the division of a given harmonic root successively by powers of 2 so as to double the number of measurable units within it.

5. The resulting *harmonic domain* (of a harmonic development) is the symmetrical region surrounding its root, in three

directions, due to exhaustion of one of the harmonic development's prime factors: (*a*) prime 3 to the left (representing division by 3); (*b*) prime 2 to the right (representing division by 2);
and (*c*) prime 5 to the base (representing division by 5).

6. The harmonic domain of symmetrical tones can only expand
(through doubling) if the prime factors of the harmonic root
locate it (as per item 2 above,) away from the left or bottom
of the matrix. This means that (*a*) it cannot expand left of the
root location until expanded to the right, through doubling;
and (*b*) it cannot expand below the root until it has expanded
above it, through doubling.

It can therefore be said that:

1. Symmetry within harmonic domains automatically clears
the submultiples needed for scales of ascending and descending intervals, between the harmonic numbers found within
that domain, enabling reuse of the same intervals by different
scales. That is, symmetry guarantees the availability of comple-
mentary reciprocal scales so that a scale and its reciprocal twin
become simultaneously available in ascending or descending
form.

2. From the perspective of modal scale formation, in which both
ascending and descending versions are required, the symmetry
found within domains becomes a necessary condition. Also, it
is not then relevant to discriminate between wavelength (string
length) and frequency (pitch), since both are guaranteed to be
co-present since they are "cleared."

3. The harmonic root $3^2 \times 5 = 45$ is the smallest root on which
all seven familiar modes can fit symmetrically with respect to
the (cornerstone) zeroth root, then requiring an interval to it of
64/45 and, by symmetry, 45/32, forming a tritone to D = root
when $64 \times 32 = 2,048$ is the cornerstone for the "cleared"
domain limit of $2^6 \times 45 = 2,880$.

THE INDIAN EXPERIENCE

It is possible that Sumeria and India had shared musical origins. The Sumerian tuning regime (see chapter 5) had reliance on the tritone and a god called Ea-Enki to give birth to the seven modes on a lyre. Ernest McClain identified India's fire god, Agni, as at the base of the tone circle (where fire rises like the phoenix) where the tritone sits opposite D on the tone circle's axis of symmetry, the eventual location of the tritone in the Sumerian system. There both semitones flip across the axis of symmetry to unseat the tonality of D itself and enter a new perspective/harmonic root/location on the mountain.

The Sri Yantra is the greatest yantra of the Hindu Tantric school, described (in the fifth to sixth century BCE) in *Shvetashvatara Upanishad*.[2] It is evolved from a bindu point into a series of interlocking upward- and downward-facing isosceles triangles, with an axis of symmetry like the tone circle.

The tone circle can be presented vertically with high D above and low D below, like a vertical monochord, conceptualizing the available scale in the Sri Yantra as having horizontals representing the note classes of the octave. My interpretation is that the horizontal lines of triangles describe the tuning operation found in the Sumerian tuning text in which each note class moves once. So, by showing the octave between two points on the same level, a set of note-class horizontals show an act of tuning from one note class to its neighbor one semitone away. Each act of tuning alternates between the top half and the bottom, in the order given in the Sumerian text.

As usual, one needs to tune a lyre in the Pythagorean manner using fifths and fourths, working in three pairs of symmetrical tunings from the octave limits, that is string 1 and string 8, using the scheme below. In theory this forms the Pythagorean heptatonic, equivalent in scale to the modern Dorian, though the final semitone gaps of E–F and B–C might be adjusted by ear to sharpen the leimma by something close to the syntonic comma of 81/80.

We saw in chapter 5 that an eight-string lyre is naturally tuned to the Dorian mode by using the largest intra-octave intervals of the fifth

Sri Yantra Presents Note Class Tuning Actions between High and Low D

All of the actions performed in the Sumerian tuning method can be described by the horizontals available within the Yantra's triangles.

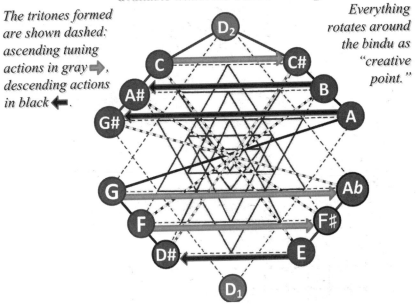

The tritones formed are shown dashed: ascending tuning actions in gray ➡, descending actions in black ⬅.

Everything rotates around the bindu as "creative point."

Figure 6.9. Sri Yantra as aide memoire to tritone tuning of modes.

and fourth, because these are the easiest intervals to tune by ear (see figure 5.5). The modern Dorian would be the Sumerian embubu tuning, containing the tritone B to F. This allows either side of the tritone to dissolve its adjacent semitone and move the scale order to a new mode. A new semitone is formed at one end from an existing tone, then still presenting a new tritone between two semitones, ready to repeat the process but then from the opposite side of the tritone. The relationship to the Sri Yantra comes from the fact that each of six inner note classes are only modified, by a semitone, once in the entire process, and there are six vertically symmetrical horizontal vertices of the triangles that make up the yantra.

One notes in figure 6.10 that opposite note classes looking across

Sri Yantra as an Aide-Memoire for Sumerian Scale Shifting

The tritones formed are shown dashed:
ascending tuning actions in gray, descending actions in black.

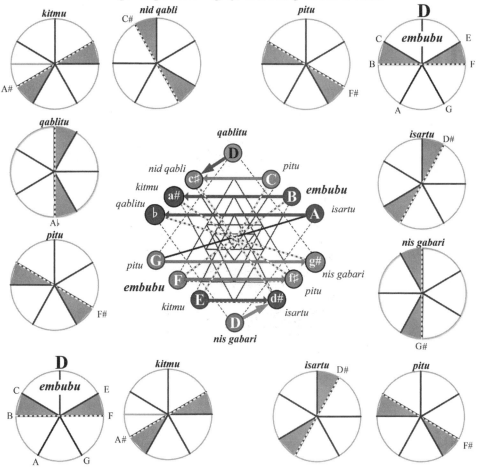

Figure 6.10. Sri Yantra with Sumerian tone circles. The yantra could have been used to visually remember how to tune between modes, as the Sumerian tuning instruction did.

the diagram intersect the bindu point of its creation so that when B is tuned down to A#, it then faces F# in the Sri Yantra, the opposite direction of tuning action, while A# is now tritone to E, which will move to D#.

PSYCHO-COSMOLOGICAL
HARMONISM IN THE VEDAS

We can now move to the inner significance given to tuning in India. The modern world is founded on objects and objective facts: we collect them, prove them, catalogue them, and even count ourselves among them. This impulse to be something is, however, but one side of reality, since nothing comes to be what it is without some interval within which it came to be. Were one to focus on such intervals (as India's system of scales did: "The seven names indicate in the first place steps, not notes"[3]), to study them *rather than* their fruits (which are tones as vibrations) then one might master *becoming* rather than be a collector of facts. This perspective has something to say to us about the purpose of learning in the modern world: that we are over-identified with the results of our own actions and unwilling to sacrifice immediate gratifications for either skill development or an investigation into the purpose of our working. So it is in the area of harmony: there is chaos until one turns to how tones arise through intervals. The Indian notion of sacrifice applied as strongly to understanding how one understands as it did to understanding the harmonic world because the rishis who wrote the Rg Veda saw their own makeup as being harmonic within their harmonic worldview, based on sound. Objects and objective facts are the preferred worldview for sight rather than sound.

> Sound in motion is the very illumination and institution of this [Rg Vedic] vision . . . intent on *becoming* that which it knows: the creator, the means of the creation, and the creation.[4]

Antonio de Nicolás, instrumental in the writing of *The Myth of Invariance,* presented the proto-Brahmins of the Vedas as being transformed according to lessons learned from the world of harmony, and here it is possible to add to harmony and self-development a further dimension of objectivity in that their gods and their yoga were probably inspired by the harmonic nature of the planetary periods. De Nicolás lived many years in India and, from the start, experienced sound as a creative force in the human sensorium, an experience normally relegated to the subconscious. A fortunate meeting with McClain led to *The Myth of Invariance,* while his own *Meditations through the Rg Veda* details the psychological import of the Rg Veda.

The harmony of the World Soul is only a necessary conceptualization of the extremely dynamic and psychological reality created as vibration but emanating from a world of relativity, which is of intervals rather than notes. There is a Zen saying: "First there was a mountain, then there was no mountain, then there was"; the harmonic mountains exist because consciousness is at work and is variable according to the perspectives adopted. I obviously agree with McClain, that harmony must have been deliberately alluded to in ancient texts, but this viewpoint does not infer that ancient harmonists had a pathetic fallacy or mere metaphor in mind. Once the planetary harmonies are correlated with the texts, it seems obvious that harmonists belonged to a significant school of ancient thought about harmony as a cosmological first cause. The psychological implications of harmony within de Nicolás's work confirms an ancient Indian tradition where cosmology was integral with self-development or yoga.

The inner world of the sensorium is where meaning can happen, but within it lies a set of rules that hides control of the meaning-making process. The thrust of the Rg Veda is to transform consciousness through dis-identification, where identification* is the creation of a god on the mountain. Limiting numbers are the limits of what can happen within them, and what can happen is a form of consciousness, a god: as in Genesis, "You shall be as gods and know good and evil."

Consciousness does not exist without questioning reality, and such questioning is suppressed when it is wholly identified with something. There has to be an *unknowing* for consciousness to arise, and I believe it was in this sense the writers of the Rg Veda proposed their system of inner sacrifices, of what they thought. According to de Nicolás there was an overall view that harmony was not a fixed absolute but a continuing

---

*The habit of identifying with things and ideas at the expense of authentic personal experience.

and dynamic relativity. And when we look at human beings there is no capacity for being able to understand everything, nor any need to. All of human life is a relativity, yet absolutes such as ideologies can capture the human relativity and turn it into a monster of identification.

And if humans are relative systems, and we know harmony is a relative system (having only one fixed *reference* tone in its body), there are only intervals. So what are intervals or modal scales or octaves? Harmony is a massive system of interrelation possible through number and without any necessary physical concerns (such as friction) until music needs to be made on an earthly instrument. To realize harmony there must be vibrations in the air on earth while, in heaven, planets roll around in orbits that do not alter over millions of years, hence vibrating through an eternal motion relative to each other. This is not to say that there are not absolute invariants but that these invariants are then applied relatively within real systems.

The planets, as materialized vibratory systems, present a nearly perpetual motion that, once conforming to the ideal world of harmony, can be maintained for long periods of time while, of course, not being the immaterial principle of musical harmony itself, based on numbers. The appearance on Earth of celestial harmony coincides with the emergence of *Homo sapiens* (see figure P1.3 on page 10), creatures capable of consciousness who have minds (and ears). In fact, harmony has more to do with subjectivity than objectivity. The planets may be harmonious, but that has little meaning without beings able to turn facts into meaning and hear harmonic intervals. It is also true that planets do not display music exactly but rather occupy locations within the framework of harmony called just intonation. If harmony is subjective, it can be objectified as a material fact while also being part of human nature through the senses and part of the human mind through number. And it is highly likely that the human mind includes that part that can receive harmony, perhaps because the heart (not the physical heart but a spiritual center) is the true organ of knowing, as the Egyptians insisted.

In evolutionary terms, the human mind is a development that has betrayed nature, as it no longer serves any obvious objective purpose. Doubtless it arose to ensure human survival, but now has taken up all available natural resources. Perhaps that is its objective purpose. The intellect can draw on an invisible resource, its own possible ways of thinking, and although humanity is preoccupied with resource management, this general-purpose nature of the mind appears linked, like harmony, to the structure of reality itself. This is most clearly seen in the creation of cosmologies—common to most religions and to recent scientific thinking.

The mind, strongly linked to language, is an instrument that can play itself, and the god in this mind is identification, over which many minds have no control. When a particular mental identification is broken, for whatever reason, something new arises that was hidden behind the identification. While this fact is in principle well-known, the power of the identification process must, like a "dragon," be consciously defeated before the "gold" on which it "sleeps" can enter the human ecology of mind—yet society as a whole is found powerless to see the problem as due to their own thinking, treating gods, myths, and cosmological doctrines as absolute rather than transitional objects.

Something about the world holds humanity in identification, leading to destructive unhappiness and lack of fulfillment. And so, the harmonic doctrine, in relating to human development, appears to have moved (since 600 BCE) to ideas of self-transformation, liberation, nirvana, self-sacrifice, avatars, saviors, and messiahs. And the harmonists of the Rg Veda seem early to have laid out how they would use the gods to do their creative work with consciousness, a far more relevant aim than following the fate determined by the gods or of kings thought possessed by the divine. In this way, the subjective nature of harmony, when established, could speak to the subjective human nature and its purpose on Earth.

The connection of the human mind with cosmology—at whatever viewpoint-defining level, everyone always has one—is a relativity that seems to orient the sense of identity within the world. And it was possible for the Vedic rishis to see access to the tritone, and other features within scales on the mountains, as presenting them with an opportunity to change consciousness, defeat identifications, and not remain trapped by them. Their model for this was associated with the need to sacrifice what is already thought, and today, the human condition regularly inflicts its process model of the world on itself as an object identity. This requires a further investigation of whether the Vedic rishis used a metaphorical language based on tuning theory or whether they were impressed that the planetary system had arisen and was actually organized that way and that, were we connected to it, it would support, if not guide, human thought.

THE ANCIENT INDIAN SCALE

Though Curt Sachs states in his 1943 *Rise of Music in the Ancient World, East and West* that "India's scales are numberless,"[5] he defines a single ancient organization of the Vedic scale, found behind the variety and variable notation of Indian syllables used to define scales. Indeed, Indian scales differed from Greek or modern scales in their use of seven syllables to indicate steps, that is, intervals, rather than notes.

The Hindus came to a division into twenty-two subintervals rather than seven intervals within an octave scale or our modern twelve semitones. However, the need for twenty-two srutis (as they are called) hinges on the ability to ring all the possible changes within the octave, given a just intonation taken to a very high limiting number. Yet, while the numeracy implied might be high, the twenty-two srutis were a masterful *simplification* of all the potentials possible to the harmonic mountains that interested the rishis and their expression of just intonation.

In figure 6.11 showing the srutis, a just diatonic scale is primarily involved, and secondarily, the Pythagorean F is used that, in tuning order F–C–G–D, requires a limit for D with a harmonic root of three cubed ($3^3 = 27$) so as to enable three steps equaling a descending interval of 16/27 to reach F. It is also therefore true that, in order to reach F on the mountain, using practical srutis, the limiting number has to be 4,320 in order to manifest the reciprocal tone B (though not used in figure 6.11). This number 4,320 is at the heart of Vedic and post-Vedic large numbers, involving the creator god Brahma and the yuga cycle* (The nine srutis of ascending and descending fourths, to

---

*Probably inspired by the precession of the equinoxes.

Figure 6.11. Curt Sachs's derivation of the twenty-two Indian srutis within the ancient Vedic scale. Its genius was to identify component intervals within tones and semitones that the most sensitive could discriminate so as to build seven scales out of a common scheme. While srutis are normally vaguely defined, a fixed system was originally intended to be made up of three srutis: syntonic commas; chromatic semitones; and the Pythagorean leimma, from which the scales found within holy mountains can be theoretically constructed.

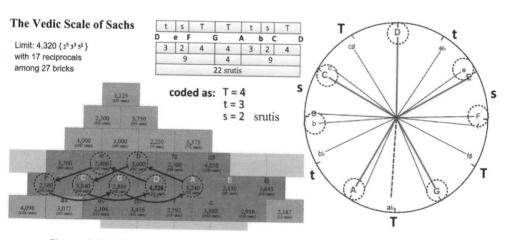

Figure 6.12. The mountain for limit 4,320, necessary to generate Vedic division of the octave. The location for D belongs to the 4,320:8,640 numbers of Vedic time periods, flood heroes, and arks. It has extra width in its central register F–C–G–D–A–E–B, which is heptatonic and offers multiple versions of the same just scales on its mountain. The scale marked on the mountain (*left*) is also marked on the tone circle (*right*), within which one can see the srutis making up tones 10/9 and 9/8 and semitone 16/15. Yet there are Pythagorean leimmas, syntonic commas, and chromatic semitones dividing these up. Unrolling the tone circle, one can clearly see how every interval within the scale is made up of two or three srutis, with values of 22, 70, and 90 cents. This indicates how the Bible and the Vedas chose different but adjacent locations for D on the mountains they developed.

G and A, match the tetrachords of ancient Greece.) And the scale is therefore asymmetrical relative to the G–A axis of symmetry (wherein the fundamental tritone of Agni resides, in-between).

Sachs relieves these mysteries by developing a substructure for this scale's adopted tones and semitones in which a few subintervals of widely varying character divide up the tones and semitones, as shown in figure 6.13, coded as *t* is just tone 10/9, *T* is Pythagorean tone 9/8, and *s* is just semitone 16/15. Obviously, the scale is based on some form of just intonation based on interval ratios with both prime 3 and prime 5.

While the t–s–T sequence appears left to right, the underlying structure of these srutis* is capable of being symmetrical since the larger tone could arrive at E and the semitone then end on f, to reach G as the lesser, just tone g; that is, in the sequence 4–2–3 srutis.

The other innovation is the implied discovery of the ratio 81/80,

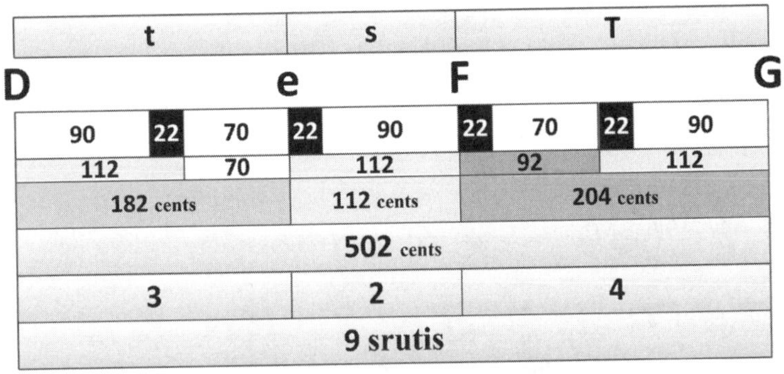

Figure 6.13. The ancient Vedic scale according to Sachs's allocations of srutis. The tabular form shows that the Indian approach was to disassemble tones and subtones into at least three components worth 22 cents (the syntonic comma of 81/80), 70 cents (the chromatic semitone of 25/24) and 90 cents (the Pythagorean leimma of 256/243), within which the scales could be realized, transposed by a fifth. The Pythagorean tone (T) and just tone (t) differ by 81/80, the syntonic comma. But then an extra unit of 22 cents is shown within all three interval types, revealing that the Pythagorean leimma can be found in all three intervals, (t, s, and T) while the chromatic semitone is within each of the tones (t and T).

---

*Srutis were never considered a fixed interval type, but probably as component intervals shown here as 22 cents (the syntonic comma), 70 (chromatic semitone) and 90 (Pythagorean leimma).

The Ancient Organization of the Vedic *Sa-grama* Scale as Derived from the Limit = 4,320

[after Curt Sachs*]

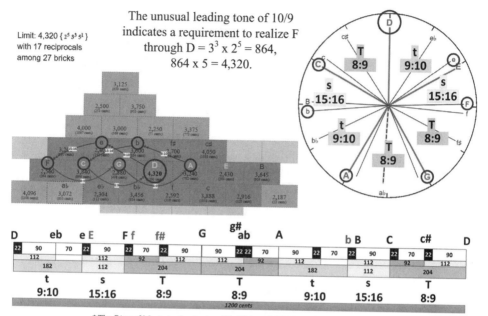

The unusual leading tone of 10/9 indicates a requirement to realize F through D = 3^3 x 2^5 = 864, 864 x 5 = 4,320.

Limit: 4,320 {$2^5 3^3 5^1$} with 17 reciprocals among 27 bricks

The Rise of Music in the Ancient World. page 166. W. W. Norton, 1943.

Figure 6.14. Cross-relation of Vedic scale with 4,320. This shows the Vedic scale was a third view of flexible tuning, additional to the mountain and tone circle, and able to arrive at the maximum flexibility in forming scales, derived from a higher harmonic root than the Bible's 45, namely, 135, from which the 4,320 and 8,640 head numbers of the Indian great yugas flowed. Indian music and cosmology were thus united.

the syntonic comma of 22 cents, which relates just and Pythagorean note equivalents in either positive or negative pitch displacement to each other.*

How this looks in the context of the whole octave can be realized by the limit 4,320, which is a just-tuning version of the Pythagorean heptatonic scale limit of 864 (864 × 5 = 4,320), in figure 6.14.

---

*The use by Sachs of modern logarithmic cents (formula: 1,200 log₂ (ratio), giving 1,200 cents to the octave) is evidently not an ancient usage but rather a useful way of accounting in the ideal sense, and is not proposed here as historical.

All of these 22-cent syntonic commas, clearly visible on our tone circles, are not used in Sachs's Vedic scale, but are available to represent scales found in all sorts of limits on harmonic mountains. But where does this take us on the mountain except onward and upward? Doubling 4,320 to 8,640 creates a fuller rhomboid than the bed of Ishtar and hence enables all seven just scales, available to limit 2,880, while also providing the fully Pythagorean scale (F–C–G–**D**–A–E–B). In *The Myth of Invariance* McClain saw the Pythagorean scale as the primary invariance, rather than modal, just scales, referencing these seven tones to "the seven rivers" or "seven tribes" of India.

The super-gods of India, such as Brahma and Indra and the system of yugas, seem to have the same limit of 4,320 or (× 2) 8,640 or (× 2) 17,280 times a number of powers of 10 to define them. McClain decodes them as shown in the accompanying box.[6]

How the Rg Veda and Other Traditions Propose Limits Based on 432,000

Hamlet's Mill by de Santillana and von Dechend powerfully connects this number within Eddic, Rg Vedic, Homeric, and Cambodian designs for the world or its end times.

In the Edda:

It is known that in the final battle of the gods, the massed legions on the side of "order" are the dead warriors, the "Einherier" who once fell in combat on earth and who have been transferred by the Valkyries to reside with Odin in Valhalla—a theme much rehearsed in heroic poetry. On the last day they issue forth to battle in martial array. Says the *Grimnismal* (23): "Five hundred gates and forty more are in the mighty building of Walhalla—eight hundred 'Einherier' come out of each one gate—on the time they go out on defence against the Wolf."

That makes 432,000 in all, a number of significance from of old. (*Hamlet's Mill*, 162)

In the Ṛg Veda:

> This number must have had a very ancient meaning, for it is also the number of syllables in the *Rigveda*. But it goes back to the basic figure 10,800, the number of stanzas in the *Rigveda* (40 syllables to a stanza) which, together with 108, occurs insistently in Indian tradition. . . . Again, 10,800 is the number of bricks of the Indian fire-altar (Agnicayana). (Ibid.)

Ernest McClain's decoding, from *The Myth of Invariance*, pages 79–81, is as follows:

A possible allusion to the Kalpa number can be heard in verses 2 and 3 of hymn 4.58:

LIMIT

So let the Brahmān hear the praise we utter.
This hath the **four**-horned Buffalo emitted.

$$4$$
$$+$$
$$3$$
$$+$$

Four are his horns, **three** are the feet that bear him;
his heads are **two,**
his hands are **seven** in number.

$$2$$
$$+$$
$$10^7 = 10,000,000$$

Bound with a triple bond the Steer roars loudly:
the mighty God hath entered in to mortals.

$$= 4,320,000,000$$

The Brahmā number can also be found in "poetic" form if it is legitimate to use some imagination:

With Horses of dusky colour stood beside me,
ten chariots, Svanaya's gift, with mares to draw them.

8 cows
_60,000 kine
_40 bay horses

Kine Numbering **sixty thousand** followed after.
Kakṣīvān gained them when the days were closing.

$$= 864,000$$

Forty bay horses of the **ten** cars' master
before a **thousand** lead the long procession.

× 10 chariots
× 1,000 lead cars
$$= 8,640,000,000$$

An earlier gift for you have I accepted
eight cows, good milkers, and three harnessed horses.
(1.126.3–5)

Homeric Greece:

> 10,800 is also the number which has been given by Heraclitus for the duration of the Aion, according to Censorius. (*Hamlet's Mill*, 162)

Babylon:

> Berossos made the Babylonian Great Year to last 432,000 years. (*Hamlet's Mill*, 162)

Cambodia:

> Shall one add Angkor [Wat] to the list? It has five gates, and to each of them leads a road, bridging over [the] water ditch which surrounds the whole place. Each of these roads is bordered by a row of huge stone figures, 108 per avenue, 54 on each side, altogether 540 statues of Deva and Asura, and each row carries a huge Naga serpent with nine heads. (*Hamlet's Mill*, 162–63)

The sruti system of India therefore provides a very comprehensive system of scales based on a limit of 8,640 but capable of describing the most significant "property classes."*

This is particularly clear when six powers of 10 can drive it upward to Brahma's number, also that of Indra's mastery over Vrtra and Marduk's number in mastering Tiamat using his flood weapon. However, the Sumerian musical systems seem a subset of the Vedic scale, and this could be a forensic sign that the similar feats of heroism were borrowed from Vedic India with only the secret of the seven just diatonic scales, not resolved as the Vedic super-scale. Alternatively, Vedic India could have taken ancient Near Eastern sources and developed these into what the rishis saw as their fuller logical conclusions,

---

*Plato likened Pythagorean and just symmetrical tones to the greater and lesser property-owning classes of Greek cities, cities then taken somewhere by D equaling a limit upon the mountain, often with special numbers to demonstrate types of government (see chapter 5).

of yugas and super-gods, and a more versatile yet practical framework for scales.

The above facts validate McClain's (and de Nicolás's) assertion that the Vedas and other texts alluding to harmony emerged from something very similar to the holy mountains in *The Myth of Invariance.*

Figure P3.1. The seven caves of Chicomoztoc, from which arose the Aztec, Olmec, and other Nahuatl-speaking peoples of Mexico. The seven tribes or rivers of the old world are here seven wombs, resembling the octaves of different modal scales, and perhaps including two who make war and sacrifice to overturn/redeem/re-create the world.

PART 3

THE WAR IN HEAVEN

The harmonious world of scales comes into being through the intervention of the prime number 5, ameliorating the problems when an octave's scale is generated only by prime number 3, within a womb (the octave) created by prime 2. While this intervention of 5 cures the large implied numbers within Pythagorean scales, an excessive intervention by 5 opens up another realm of giantism among the gods, in the vertical dimension of McClain's holy mountains. These disharmonies also need to be reconciled, from a planetary perspective, since the inner solar system forms a "flying serpent" raised up by the fourth power of 5 (625) on such mountains.

This problem from "on high" was never fully stated in the ancient world, though the Babylonian god Enlil, who ordered the flood, and the super-gods who saved mankind from the giantism of prime 3 were themselves highly elevated by prime 5 on the mountain. There are problems for mankind in such harmonic heights, these only hinted at by obscure stories: the flying serpent includes the eclipse year as one of its tonal values. Eclipses galvanized ancient populations, their rulers, and priests. It is possible that one or more unallocated gods might have represented the eclipse cycle—mistaken today as solar or lunar deities. For example, flood heroes have little to do after their great work but to become sacrificial fertility gods, who hover over the miraculous mysteries of resurrection, a halfway region within the octave's tone circle, centered on the tritone and square root of the octave's 2, of doubling (chapter 7).

Regarding sacrifices, the great unmentionable is human sacrifice, which appears connected to eclipses. Missing evidence of the flying serpent in the Old World may be due to censorship after human sacrifice to appease the gods was suppressed. The proposed sacrifice of Isaac by Abraham, an ancient Near Eastern norm, finds the Lord God testing Abraham's obedience perhaps, but it is more likely the story indicated human sacrifice was unnecessary—ejecting it from a religion without gods.

The demise of human sacrifice may coincide with the sudden collapse of the Bronze Age in the eastern Mediterranean (1200 BCE). In the previous centuries the extraordinary Olmec civilization had become established in southern Mexico, at the end of the Mediterranean and Atlantic sea currents and trade winds. They provide strong evidence, ignored for various reasons, of a transmission of ancient Near Eastern metrology, mythology, geometry, architecture, megalithism, and numerical tuning theory.

Only the Olmec seem to have fully resolved the flying serpent, calling it after the hero who transmitted the ancient sciences to them, Quetzalcoatl—the Feathered Serpent and their name for the planet Venus, who forms the head of that serpent's body. The Olmec arose between 1500 and 1200 BCE and had a civil culture involving, if not centered on, human sacrifice. They innovated the *tzolkin,* a sacred calendar of 260 days, unheard of in the Old World. The tzolkin lies at the heart of the Feathered Serpent and beside it lies the eclipse year. The harmonic limit of 2,880, required to resolve the eclipse year, was captured in a stone relief at Chalcatzingo (600 BCE). Later, the Olmec city of Teotihuacan was built using harmonic relationships measured in megalithic yards, between the Pyramid of the Sun and the Pyramid of Quetzalcoatl, this based on Adam's limit of 1,440, half of 2,880 (chapter 8).

Chapter 9 collects the locations of the flying serpent, Quetzalcoatl (Mercury, eclipse, tzolkin, Mars, and Venus) and of the lower serpent (Jupiter, Saturn, and the lunar year) into a matrix limited by the number given for YHWH (and Apollo,) 60 to the power of five or 777,600,000. The final chapter then solves, as well as is likely possible, a mystery left by Ernest McClain: that of "finding the lost sheep." In this, God on the Mountain (Osiris) combines with YHWH/Apollo to create their incarnation or messiah, having all of the Pythagorean and (simpler) just tone sets surrounding 777,600,000 but "on Earth." These locations on the astro-harmonic matrix are found to be three ancient calendars, based on 360/361 days (Near Eastern), 364 days (Near Eastern) and 365 days (Egypt and Mexico): indicating that ancient calendars had hidden harmonic origins (chapter 10).

The overall picture is one of dissonance between excessive powers of *either* 3 or 5, these only finding a perfect balance in just intonation's judicious combination of these primes. The smallest limit for this is 1,440 (and 2,880 for seven scales), which then fits the planetary world seen from Earth, while the overall structure of the heavens is balanced within YHWH/Apollo's limit of 777,600,000. The Earth and the moon have placed the planets (and their associated calendars) into an order in which human beings can, perhaps more easily, find their own salvation.

7

Gilgamesh Kills the Stone Men

The story of Gilgamesh hinges around his relationship with Enkidu; they became fast friends after proving to be equal wrestlers.[1] Their image was closely allied to the zodiacal twins we now call Gemini, of whom it was said that Castor was semidivine and Pollux divine. Similarly, one finds that Gilgamesh was two-thirds divine while Enkidu only one-third divine—a more harmonic way of putting it. A reference to Gemini at this date would probably refer to the recent precessional period, when Gemini was the equinoctial constellation for spring— roughly between 6000 to 4000 BCE—which archaeology tells us was a period of Neolithic experimentation, when partly agrarian lifestyles were evolving.

For Gilgamesh's walled city of Uruk around 4320 BCE, the equinoctial sun had only recently moved out of Gemini and into the sign of the Bull (Taurus). Enkidu's style of life was therefore regressive, that of a "wild man," hunting and "living with" animals. In Europe, huge henge monuments were replaced by stone monuments such as stone circles and standing stones, but in Mesopotamia the changed age manifested cities, necessarily supported by agriculture. Thus, these twins were emblematic of a contradiction between adjacent ages, and as a way of understanding progress, when progress was seen as being due to changes in the cosmic world, as the will of the gods. These

Figure 7.1. Statue of Gilgamesh. Photo by Urban (CC-by-sa 3.0)

Figure 7.2. The constellation Gemini the Twins goes back at least to the Babylonians, who knew the group as the Great Twins—a reference to Gilgamesh and Enkidu from the epic of Gilgamesh. This image is "Gemini," by Sidney Hall (1825). Restored by Adam Cuerden.

"twins" interacting were the gods as heroes, initiating historic changes and showing human beings, through their divine proportions, how to view the present as changed.

CUTTING DOWN
A MONSTER'S TREE

As soon as Gilgamesh and Enkidu became true friends, "Time passed quickly. Gilgamesh said, 'Now we must travel to the Cedar Forest, where the fierce monster Humbaba lives. We must kill him and drive out evil from the world.'"[2]

For a third-millennium Mesopotamian audience, there was no need to specify the location of the Cedar Forest, since cedars came from Lebanon, from where the headwaters of the Euphrates, one of the *twin* rivers of Mesopotamia, came. Lebanon was north of west and on the eastern coast of the Mediterranean basin. The two "brothers in arms"

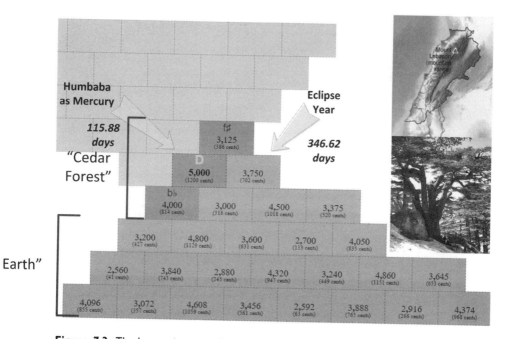

Figure 7.3. The harmonic matrix for the limiting number 5,000, if a harmonic metaphor for the journey of Gilgamesh and Enkidu was intended. *Top right,* image by Dr. Brains (CC-by-sa 3.0); *bottom right,* image from Benutzer: Mpeylo (CC-by-sa 3.0).

set off, traveling 1,000 miles every three days and taking fifteen days to cover 5,000 miles.*

The number 5,000 can be decomposed into factors of the primes 2, 3, and 5, as 8 × 625 (which is 2 to the third power *times* 5 to the fourth). In the astro-harmonic matrix the limiting number 5,000 corresponds to the planet Mercury.†

The monster Humbaba had been placed there by Enlil, the god of harmonic heights, whose powers of 5 raise the harmonic mountain up, and he was there as well to guard the home of the gods in the highland region of the Cedar Forest around Mount Lebanon. It will later be shown that the planetary gods of the inner solar system (Mercury, Venus, and Mars) have all been raised up *dis-harmonically* in the table of powers by four powers of 5, or 625, from the bottom rows where the lunar year and the Jupiter and Saturn synods relate to each other with the tones and semitones of just intonation. The height of the cedars is symbolic therefore of the harmonic height of these planetary gods above human life. Humbaba-Mercury is also guarding the eclipse year, just a musical fifth apart from Mercury. Eclipses were a powerful symbol for change: the sun or the moon being eclipsed stands for how the Age of Taurus had eclipsed the Age of Gemini.

Mercury can only be seen when apart from the sun, either rising before sunrise in the east or setting after sunset in the west. From Uruk and surrounding cities Mercury would be seen at times in the west and at other times in the east, twisting and turning in its proximity to the sun in an intestinal fashion as per the monstrous iconography of Humbaba in the ancient Near East. Humbaba was presented as protecting the sun and its meaning, and the solar hero seeks to find that meaning and alter it, so as to "drive out evil from the world," which accumulates. Enkidu helps Gilgamesh cut off Humbaba's head, which becomes a trophy, prefiguring (by Archaic Greek times) the decapitation of Medusa by Perseus, who was then

---

*The actual distance to Mount Lebanon would be about 666 English miles.
†Since the location of 625 divided by 1/160 of the lunar month equals 115.354 days, while Mercury has a synod of 115.88 days.

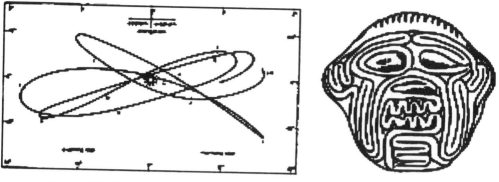

Figure 7.4. The equation (*left*) between Mercury's erratic orbit around the sun and (*right*) the god Humbaba, whose motions resembled the intestines, often used for augury. Sippar, southern Iraq, about 1800–1600 BCE.

more clearly a solar hero.* The change of heavenly framework, and not the gods themselves, was the means to "drive out evil" when an existing order was no longer ordained, and no matter what its merits had been.

Mercury, portrayed as Humbaba, was the root of a "tree of meanings" that constituted how a current age saw its world. Without toppling that tree, it is impossible to culturally sustain an alternative view of the world. The sky gods must be forced by a hero doing something new, which is required for change to happen.

KILLING THE BULL OF HEAVEN

When the heroes returned to Uruk, on cedar logs floated downstream to Uruk, Inanna/Ishtar† moves to seduce Gilgamesh (to whom she is

---

*Perseus was transformed in the Taurus Mountains above Lebanon into Mithras, who, as slayer of the Bull, would change the course of the solar year from one equinoctial gate (Taurus, following Gemini) into the next (Aries, following Taurus), as with eclipse nodes causing eclipses, due to a backward sliding of the equinoctial points through the zodiac. See David Ulansey, *The Origins of the Mithraic Mysteries.*

†The Sumerian name for Venus was Inanna, but since the Akkadian and Old Babylonian empires inherited most of the Sumerian pantheon her name became the Akkadian Ishtar, which is the name found on cuneiform tablets that tell the story of Gilgamesh.

related via Enki, sharing maternal lines), but he declines her, pointing out the fate of her past lovers such as his cousin (and previous king of Uruk), Dumuzi.

Furious, Ishtar calls on Enlil (her father and air god of the city Nippur) to unleash the Bull of Heaven, a bull perhaps identifiable with the new equinoctial constellation Taurus. But the Bull is then killed by Enkidu, who holds its tail while Gilgamesh slices between its shoulders. Enkidu cuts off its right thigh and throws it in Ishtar's face (who takes it off and enshrines it in the heavens). The Thigh is the "plough" part of Ursa Major's brightest stars, in India called Seven Sages, a constellation considered the rudder of the celestial ship. These seven stars had been on the equinoctial meridian but were now displaced by whatever powers cause a new zodiacal age to come about and shift the equinoctial meridian, this seen in the precession of the Earth's axis, its North Pole.* This projection by Enkidu of the Bull of Heaven's thigh (symbolic of the end of the Age of Gemini) caused Ishtar and the gods to kill Stone Age Enkidu through a weeklong loss of vitality—disturbing to Gilgamesh—through which the gods withdrew their support from the wild men.

A SEARCH FOR RESURRECTION

Gilgamesh cannot bear his friend's demise, seeking to bring him back to life. To do this he must find the one mortal who had escaped death, this time by going to the "twin peaks" atop the world, where he meets a scorpion man and woman, the latter telling him how to locate the place where the sun travels into the Earth after its "death"—its setting—in the west via a tunnel only accessible from the entrance of the sun's

---

*Precession arises because Earth's equator deviates from the path of the sun except at the equinoctial points of spring and autumn, a deviation called "the separation of the world parents" of earth and sky. There is a similar deviation of the lunar orbit from the path of the sun, and this causes eclipses when the sun arrives at a lunar node, which *resembles* the form of the precession of the equinoxes. These errant "points" appear as sources of change on Earth and have the common characteristic of deviance between two of the pathways in heaven, the celestial equator, the sun's path, and the moon's path.

dawn, in the east. Scorpio at the time was the autumnal equinoctial constellation, standing opposite in the zodiac to the vernal point in Taurus.

The sympathetic scorpion woman recognizes Gilgamesh's two-thirds divine constitution and directs him to hurry through the equinoctial tunnel in "twelve hours" before the autumn sun passes through it again. Entering the tunnel, he arrives at its western end just before the sun arrives at the autumn equinox. He has then emerged into a divine landscape, with flowers of precious stones above a large ocean. He descends to a tavern where the divine barmaid Siduri "dwells by the edge of the sea." She tells him "there has never been a crossing" of the ocean because of the Waters of Death. But she also advises him of a mystery figure named Urshanabi, the boatman of Utnapishtim, whom he seeks. The boatman is a priest of Enki, god of the waters and of wisdom and knows how the dead might be brought back to life.

By way of explanation, it was the god Enki who had warned Utnapishtim of Enlil's plan, as leader of the gods, to destroy humanity with a flood (see chapter 7). This warning caused Utnapishtim to build an ark. Floods were an ancient symbol for cultural changes between ages, while the "sea of harmonic numbers" (called waters) could be a flood in an outbreak of giant inharmonic numbers (chapters 2 and 3). Thus the matter of harmony in human affairs was being conflated with harmonic number theory, originally derived from astronomy (chapter 1).

Unknown to Gilgamesh, Urshanabi could take him across the infinite ocean because he crewed his boat with stone men who could touch the waters and live. Yet, as soon as Gilgamesh sees the stone men, he smashes them and throws them into the sea,* unaware of their needed role in navigating to the eternal world to meet Utnapishtim, the keeper of the equinoctial junction for a previous age, and the one who can advise Gilgamesh on whether Enkidu can be brought back to life.

---

*Is it not likely that smashing stone men alludes to smashing their monuments, which helped the stone men navigate the Waters of Death, if those waters were astronomical?

Fortunately, Urshanabi has a further harmonic solution for navigating the Waters of Death, not requiring stone men.

One then needs to carefully connect the boatman Urshanabi with

1. An ark-building Utnapishtim, whose boatman he is;
2. The god Enki, who saved mankind from the flood by warning Utnapishtim;
3. Enkidu (Enki-du), whom the gods had just killed at Ishtar's request and who represents a Stone Age hunter;
4. The stone men, who could not be killed by the Waters of Death; and
5. Gilgamesh, who loves Enkidu but summarily dispatched the stone men.

Having no stone men, Urshanabi said the other way to safely traverse the thin current of Death was to make three hundred poles, each one hundred feet long, and use these to punt across the shallows of these waters.

> [Following instructions,] Gilgamesh went deep into the forest, cut down three hundred punting poles, each a hundred feet long, he stripped them, made grips and brought them to Urshanabi the boatman. . . . Urshanabi said, "Now be careful, take up the first pole, push us forwards, and do not touch the Waters of Death." . . . When all three hundred poles had been used, Gilgamesh took Urshanabi's robe. He held it as a sail, with both arms extended, and the little boat moved on towards the shore.[3]

In figure 7.5, 300 × 100 (the depth of the bottom three rows) gives (if one wishes) a limit of 30,000, and this provides an image (if one wishes) of a boat where the handle of each punting pole lies in the waters 100 feet above the sea (288 × 100). Gilgamesh holds up his shirt atop the matrix (at 15,625) to guide the boat to shore, both these tones propelling the enneadic (3^2 making 9) boat. The limit (D = 30,000) can be located within the astro-harmonic matrix, as the position of the (luni-

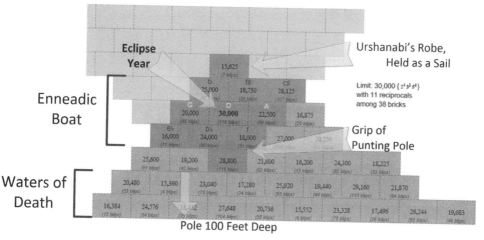

Figure 7.5. Gilgamesh's arms appear to be the chromatic semitones of 25/24 between b and b♭ and f# and f (and why not G as Gilgamesh and Urshanabi as A, while D is their common purpose).

solar) eclipse year, on the upper register of the gods (row 5), between Mercury and the tzolkin period of 260 days.* There are parallels here with Nineveh, said in the Bible to have a population of 120,000 people (Jonah 4:11), featured in the story of Jonah (see chapter 9), in that 30,000 × 4 = 120,000. Utnapishtim was therefore the original Noah, because this very story was altered to suit the compilation of the Bible in 600 BCE.

Enlil was enraged that Utnapishtim and Enki had preserved humanity from his flood, but soon realized that saving some men had been good, and he had rewarded Utnapishtim and his wife with immortality. However, in talking to Gilgamesh about immortality, Utnapishtim emphasized immortality's problems rather than its virtues, comments Gilgamesh was not keen to hear. Nonetheless, Gilgamesh is dispatched with Urshanabi, both never to return. Urshanabi is a real

---

*The Mexican tzolkin of 260 days, unrecorded in Old World calendars, was either suppressed or not understood by the ancient world. The Olmec/Mayan supplementary series, Adam's 18-lunar-month periodicity, elevated by the divine diesis (of 125/128), reaches the heart of the "feathered serpent," the tzolkin, whose head was Venus-Quetzalcoatl (see chapter 8).

prize, for he carries knowledge of the celestial region and first shows Gilgamesh where one could find a plant of eternal youth at the bottom of the ocean. But once the plant is recovered, Gilgamesh manages to lose this life-preserving plant to a serpent. Arriving back at Uruk, Urshanabi checks that its walls were indeed established by the seven stars that are also known as the Thigh (as noted above, the plough part of Ursa Major thrown by Enkidu), confirming Urshanabi a specialist in establishing the *me*, or measure, given earlier by Enki to Ishtar, who established that me in Uruk.[4]

It therefore seems the me is the thigh that was thrown into Ishtar's face by Enkidu, given to Ishtar by Enki, god of the waters and wisdom. Actually the leg of a stag, it was the idealized prey of the hunter while also being a symbol of sacred knowledge and the search for it. The flood savior Gilgamesh witnesses the tragic death of an old world (the world of Enkidu) while repopulating a new world of ideas and values (the city states of Sumeria). Enkidu symbolized the end of his world by his casting of the thigh (emblematic of his failing age) into a new position beyond the equinoctial sun—after which he loses vitality. The epic dramatized the onset of a new world order through Gilgamesh's visit to the new crossing point of the two celestial rivers—equatorial and ecliptic—and his conscription of Utnapishtim's "assistant," who transfers the knowledge of the stone men to Sumeria, where it is already enshrined into the walls of Uruk by Ishtar, that is, the planet Venus.

The tunnel of the Scorpio people was located at the autumn equinox around 2,700 BCE, and Gilgamesh's journey is from the spring equinox (birth of the year) to the autumn equinox (death of the year). Utnapishtim lives forever at this "place," the confluence of the equatorial and ecliptic "streams," where the changes in the sun's position relative to the zodiacal constellations are easily noticed in changes to the heliacal rising of stars and constellations. The constellation in which the spring equinox sunrise occurs moves slowly but continually upward until a new constellation defines the equinox for an age. The zodiac was an explanatory framework within which the agrarian and city-building cultures of Sumeria, Akkadia, and Old Babylonia had come to understand this displacement of the wild hunters, subsistence farmers, and the stone men (who had built astronomical megaliths) *as a natural process,* from whom they had inherited the nomad's husbandry, the skills of farming, and the stone men's me, or measures.

Before Gilgamesh, only the megalithic stone people had been able to cross (according to their religion) the Waters of Death, yet Gilgamesh instinctively destroys the stone men when encountered, since they were the symbol of what we now call the Terminal Stone Age, then they were replaced by the "servant of Enki, boatman of Utnapishtim," an urbane version of an only recent knowledge of astro-harmonics, rehoused in the ancient Near Eastern civilization of Sumeria.

OSIRIS AS RESURRECTED FERTILITY GOD

After Marduk's great victory (chapter 2), and long before his ziggurat was built, he seems to have been retired to live on as a popular fertility god whose celebrations were very important in the calendar of Mesopotamia. As in Egypt, the people of the Mesopotamian plains relied heavily on the land's fertility to support their level of civilization; populations concentrated around cities, where specialist crafts and arts could be developed. What happened to Marduk after being a super-god was amplified in creating the Egyptian god Osiris, who in death was resurrected every year in the fertility of the Nile valley. A fertility god *must die* to be reborn—a characteristic perhaps of tritone godheads who work at the highest level of cosmic creativity yet run parallel with how the human mind comes to think of its own creativity in understanding the world and defining the human purpose.

A numerical connection can be found between Marduk and Osiris in a line drawing in the *Egyptian Book of the Dead* made for a scribe called Hunefer, of Osiris seated on his throne (figure 7.6). Behind him is Isis, the wife who resurrected him after his brother Set had locked him in a coffin, made to fit just him, rather like the Cinderella story and her shoe only with more sinister intent. It is his throne that concerns us here for, in the ancient world, iconography could include objects that stood for numbers employing a number of different counters used for doing arithmetic. And harmonic calculation was naturally displacing geometric metrology in the ancient Near East, the counting of numbers having become useful for mundane recording of city and state trade, taxation, and inventories. As I would have it, and as Budge stated it, Osiris is sitting on a number containing seven powers of prime 5, the number of powers of 5 found in Marduk's number.

In figure 7.6, the upper seven bands within his throne are separated from the lower eight bands by the inclusion of a rectangle whose height relative to its width (as drawn) is the square root of 2, which harmonically is the tritone. If the lower bands represent the powers of 2, then the total number represented was $5^7 \times 2^8 = 20,000,000$: a type of harmonic divinity that Ernest McClain called "God on the Mountain." This sort of harmonic god is like Enlil = 50, the Sumerian chief of the gods, at the top of the octave for 30:60. Osiris stands far above Enlil on the mountain for 20,000,000, at the same height as Marduk = 8,640,000,000 but, like Enlil = 50, denuded of any of the powers of 3 that create the basis for a Pythagorean scale, that is, for musical coherence.*

The throne of Osiris sits on another rectangle filled in by vertical diagonal wave patterns (alternate chevrons) having four levels of hatching that I believe point to the bottom four rows, the "waters" of his harmonic mountain out of which two leaf-bearing stalks and a central papyrus flower emanate. The right stalk, emerging from within the waviness on the second "wave," is the moon as lunar year whose harmonic root is 15 in the planetary matrix. The left stalk is the cornerstone Saturn, sometimes equated with Set, who killed Osiris and cut him up, a brotherly intrigue echoing the early dynastic rivalries between Lower and Upper Egypt in the first century of the third millennium, in which pharaohs of both Set and Osiris appeared to rule. The lunar year is probably Nephthys, first married to Set then Osiris to become co-supporter in the tableau alongside Isis, generally regarded to be Ishtar-Venus, who, in the ancient world, was not yet reduced to an Aphrodite-Venus.

When the synod of Venus is analyzed within the time frame of eighteen lunar months = the limit 1,440 (the key limit for the astronomical

---

*Just intonation merely improves on the Pythagorean scale and relies on still using at least two or four of its tones (see chapter 6).

Figure 7.6. Osiris judging the dead, on a throne with harmonic implications, surrounded by iconography that harmonic numbers might help explain. From *Osiris, Judge of the Dead* (papyrus, Egypt, ca. 1275 BCE) in Budge, 1911.

matrix explicit in the New World, see chapter 8), Isis is found to correspond with the fourth power of 60 = 12,960,000. She is a development of the path of Anu = 60,* who shares the lunar year's harmonic root of 15 (Nephthys), times 4. Isis's power of 5 is therefore 5^4 = 625. This represents the lowest level of reciprocal tonality within the harmonic mountain of Osiris, a location that, in planetary terms, is occupied by Thoth, that is, the synod of the planet Mercury. This then places the four rows below the throne of Osiris as his pediment, which I believe is also the coffin within which his brother locked him.

---

*Anu = 60 is the first harmonic god who carried off "heaven" along his path of powers of 60.

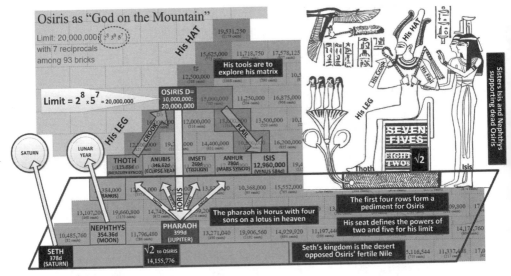

Figure 7.7. Composite interpretation comparing harmonic features of Osiris as 20,000,000 "God on the Mountain" with his iconography. Isis is supporting Osiris as the other side of a heavenly pyramid whose base is Thoth-Mercury, his lowest darkened brick. The base of this pyramid is the flying serpent of inner planets and time periods that the Olmec called the Feathered Serpent (see chapter 8). This Egyptian serpent has been lost in all but name, along with the harmonic doctrine of Egypt.

The four supporters of Osiris, emerging from a "lotus," could be a direct transcription of the four faces of Brahma, who emerges from the navel of Vishnu to create the world and, one of these being Thoth-Mercury, identifies the "flying serpent" of what we would call the inner solar system with Isis-Venus at its head. These, plus Isis herself, then form a pyramid to the base of the tall hat of Osiris, a hat* out of which the Nile probably flowed. Isis is at the right supporting Osiris, wrapped in his funerary wrappings and hence single legged, his form extends along the left-hand "darkened bricks" of his lifeless body. The synod of Jupiter lies beneath him, and below that lies a brick "on the earth" *exactly* like Marduk's, in having a value that is √2 of his D.

Pharaoh is the Horus god-king, and I propose that the √2 beneath

---

*Which in India could be Shiva's hair, the constellation Orion from which Eridanus flows.

the pharaoh is the Nile's fertility, placed centrally within a heptatonic in which the ghost of Osiris brings fertility to the Nile valley. As such the four supporters are control over knowledge (Thoth-Mercury), death (Anubis-eclipse year), a lost sacred calendar (Imseti-tzolkin), and war (Anhur-Mars). It is this planetary matrix that reappeared in the preclassical Olmec world (see chapter 8), possibly due to the twelfth-century BCE Bronze Age collapse affecting Egypt* and most other parts of the eastern Mediterranean.

Perhaps most interesting are Osiris's familiar tools, the crook and the flail. I believe the flail represents the procedure originally used to discover Osiris, who looks down on the flying serpent of the inner planets, to form a pyramid with Thoth and Isis as its base. The flail travels (arithmetically) up the slopes of prime 5 and can then look down on a negative descent to calculate the ratios of the different bricks to the eighteen-month periodicity of the planetary matrix. Large numbers can be desiccated to their factors of 3 and 5, symbolized as desiccation during mummification after death, and made to travel in the lands of 3 and 5 when matching the interval ratio known to exist between planetary synods and 3/2 lunar years. The crook represents the recovery of these tonal sheep, by traveling horizontally using prime 3. This would confirm the attitude that prime 2 is irrelevant in locating tones within a given octave because two is not generative inside its own creation of the octave, though enabling of harmonic ratios found within it. Two is supportive within octaves, as with the role of female figures Isis (as Venus) and Nephthys (as moon), as the number of powers of 2 within an octave's limit.

The whole planetary scheme is revealed when Horus is seen in his manifestation called Horus-Ra in which the sun is seen emerging from a *djed* pillar topped with four discs on which an ankh, the symbol still used for Venus, has arms placing the sun above her. Isis sits on top of the four-level pediment of the djed ($5^4 = 625$), but above her is a period of 5/8 of her synod of 584 days, the practical year of 365 days, which is 5×73 days long.

---

*"After apparently surviving for a while, the Egyptian Empire collapsed in the mid-twelfth century BC." Wikipedia, "Late Bronze Age Collapse"

Figure 7.8. The creation of Horus-Ra from out of an ankh
with female arms atop a djed. From Budge, 1899

It appears the flood of Marduk—the letting loose the waters of
heaven and causing a flood—became the life-giving water of rivers, necessary for civilization to support itself in the ancient Near East. Osiris
shows the identification with the Nile and his resurrection in the god-
king pharaoh and the Nile in seasonal flooding, giving fertile silting
and moisture. But Set always attends his brother's miracle, as the desert
lies within yards of the riverbanks.

8

Quetzalcoatl's
Brave New World

Pursuing our theme of astro-harmonic knowledge in the ancient world, the Olmec of central Mexico appear to have demonstrated it from 1200 BCE onward. The norms of their Mesoamerican culture would flow on into the Mayan and Aztec civilizations, just as the Sumerian norms had flowed on into the Akkadian and Babylonian civilizations of Mesopotamia. The Olmec are classified as the preclassical Mayan period that, in the form of the classical mega-city Teotihuacan, remained politically active into the classical Mayan period. The Olmec are the point of origin of the advanced astronomical and calendric system in the Americas. Did the Olmec arrive at their advanced knowledge and skills themselves, or did these arrive from the Old World, whether from the eastern Mediterranean or the west, from Asia?

The academic consensus is that the Olmec developed everything themselves, that is, in isolation. But the likelihood of an identical metrology arising without contact with the Old World is zero, since metrology depends on standard root lengths. Once an Old World measure is found in the New World it certainly was not locally sourced. To avoid accepting this metrological fact, proving diffusion rather than isolation in the Old World, academics favor measures being ad hoc lengths generated by dimensions of the human body, an anthropocentric and nontechnical theory, which is then supported by indigenous peoples and local governments who want any past genius to have been homegrown and on a par with the Old World civilizations. However, I believe it obvious that the Olmec were given important parts of their high civilization through contact with the ancient world via the Atlantic, or possibly the Pacific as well.

Olmec metrology, astronomical harmonism, buildings, and iconography are not just parallel to those of Old World cultures, but in many cases invariants can be found for which a precursor civilization would have had to exist.

THE ORIGINS OF OLMEC ASTRO-HARMONISM

I first encountered astro-harmonism in a monument (built ca. 600 BCE) at Chalcatzingo in central Mexico. In addition to the obscure astronomy behind counting time in 260-day cycles, the Olmec had an advanced harmonic knowledge like that described in previous chapters, and their unique iconography linked the 260-day tzolkin to the invariant time period of the eclipse year. As in the ancient Near East, harmonic knowledge of time periods was informing their religious ideas, and they built pyramids and used measures (at Teotihuacan in particular) found in ancient world structures and developed in the megalithic era for conducting an astronomy of time periods.

This "ancient" metrology is of such a specific and accurate genus it could only have been standardized in one place and at one time. This refutes the idea that the Olmec had created all of their cultural norms from scratch. In their use of metrology the Olmec must have been initiated into its use by specialists from the ancient world just as, in fact, the Sumerians and Egyptians had probably been initiated into metrology by megalith-building specialists who had originally developed metrology to measure celestial time periods but who unexpectedly deduced that the planets were harmonious to the moon. By 1200 BCE there were no megalithic peoples to directly tutor the Olmec in metrology, yet that knowledge was well established in the ancient Near East.

Olmec religious ideas seem congruent with the ancient Near East, where the harmonic knowledge behind the super-god myths had evolved toward the "serpent in the sky" as the Egyptians had it. This became the feathered serpent of Quetzalcoatl, a key Olmec concept inherited by the later Mayan and Aztec civilizations. Quetzalcoatl functioned within the Mexican origin myth that claimed their cultural life had been enriched by a person they called Quetzalcoatl or Feathered Serpent. This name was also used for the planet Venus, whose synod leads the flying serpent, thus characterizing it. The Olmec built megalithic constructions (pyramids, stelae, giant heads), and they took megalithic day counting to new heights by forming a calendar on a grand scale of thousands of years (eventually the Mayan Long Count), surpassing anything known in the ancient world, but then resembling the large limiting numbers used within the harmonism in the Old World. A case can be made for Olmec influences having included India, China, and Japan; that would depend on the Pacific's twin "conveyors" (trade currents and winds) rather than the Atlantic's sole route, from northern Africa to the Caribbean. But the Olmec myth of contact points to the east, like the morning star of Venus. The eastern Mediterranean, across the Atlantic, had the necessary metrology, pyramid building, and harmonic knowledge to have been the primary influence.

Academics have sealed off the Americas from any historical contacts with the rest of the Old World, but when all of the Olmec numerical sciences and norms are compared to the ancient world, it is obvious that diffusion occurred. The Olmec built pyramids, planned cities, developed a system of writing, and had advanced graphic arts. Also, around the moment of early Olmec city building, the Old World was in turmoil due to climate change and migration—the Late Bronze Age collapse recorded best by Egyptian records but seen in widespread devastation of settlements. Initiates of an astro-harmonic school may well have sought to preserve their knowledge away from the decline of ancient empires and the influx of new tribes from the north.

The first Olmec cities of San Lorenzo and La Venta were built circa 1500 and 1200 BCE, respectively, two fertile river valleys, each river flowing north into the Gulf of Mexico at the isthmus between North and South America. The indigenous people were ripe for the agricultural development that had supported the Old World civilizations, with the synthesis of an ideal triple crop of maize, beans, and squash, leading to the worship of corn gods. The fertility of river basins and double growing seasons allowed these "three sisters" to provide the surpluses necessary for city building and a social hierarchy. San Lorenzo was built on a vast man-made plateau where colossal spherical heads, a pyramid, and finely carved jade burials belonging to an astronomically profound jaguar (*nagual*) cult have been found. Ten thousand people are thought to have resided in a stratified society. One theory is that it might have been a tentative development, in that the sacred forms that would emerge in later Olmec, Mayan, and Aztec cities were not yet fully expressed.

A number of common Olmec features were already present in San Lorenzo:

1. Large boulders, thought to represent rulers or gods, were carved out of basalt quarried 40 kilometers to the north.
2. The existence of a jaguar (nagual) cult of metamorphosis between human and animal forms.
3. The presence of stone altars, such as a stylized jaguar mouth within which a form-changing priest is sometimes looking out, beneath a diagonal cross motif (Monument 4).
4. A ritual ball game was played on a court with rubber balls.
5. Water sacrifice of wooden effigies with faces like the faces on basalt boulders, and of rubber gaming balls, thrown into a lake.

San Lorenzo seems to have been suddenly destroyed, with civic violence or in favor of La Venta—a sacred city of about the same population. La Venta contained many additional Olmec features; the site was oriented north-south, becoming the prototype for the later Mayan city-states, containing a pyramid (made of 100,000 cubic meters of clay,) a more detailed ball court, and many ancillary buildings and complexes, all for the same number of people as San Lorenzo. It is significant that a type of rock called serpentine was brought east from the Pacific coast, 4,000 tons of which were used in front of the pyramid to form a voluminous exotic pavement employing a standardized jaguar-inspired pattern of upright and cardinally placed axe-head shapes.

Balls, boulders, and jaguars are connected within Mayan astronomy, which explained an eclipse as the moon or sun being eaten by a jaguar, because the jaguar (nagual) uniquely takes the skull of its prey in its mouth in order to crush it. Ants were another explanation for eclipses, perhaps leaf cutters. According to Susan Milbrath, "The most common explanation for eclipses among the Maya is that there is an animal devouring the sun or moon. . . . Sometimes the animal is a jaguar or a *tigre*, but more often it is an ant."[1] The diagonal-cross motif is indicative of the moon's orbit crossing the path of the sun on one of its "lunar nodes," which were called draconic in the Old World since dragons were thought to swallow the sun or moon at these points. These crossing points can only be tracked by understanding how they move within the time counting of a calendar. But the eclipse year is a part of the body of the Feathered Serpent, a cycle of musical fifths raised up (hence flying) by a minor diesis (125/128), so the eclipse year is to be seen harmonically flying above the lunar year. The Olmec iconography can be loosely explained by their astronomy and calendar, but more exactly explained as *harmonically* locating the jaguar's eclipses above the heads of the "priests" sitting within altars. It is likely their cult of metamorphosis into the jaguar was connected to the power of eclipses as harbingers of a harmonic metamorphosis connected to the appearance of the tritone "twins," connected to the corn-god twins who die and are resurrected like northern Europe's own Green Man and John Barleycorn.

Evidence of this harmonic location for eclipses can be found at Chalcatzingo (figure 8.1), a small Olmec city located beneath twin peaks on the trading route to La Venta. It was probably on the supply route for the serpentine stones of La Venta. Its many special monuments include a bas-relief called El Rey, which assumes the figure shown is a male king, yet she is wearing female clothes and so should be called La Reina.* The city is a small but beautiful expression of Olmec architecture and

---

*"The local name is spelled 'El Rey' ('the king')—which assumes that this is a male; however, I've argued that this a female ruler. The identification as a female is based on the fact that she wears a skirt, and has long hair. So this relief should be called 'La Reina' rather than 'El Rey.'" Michael D. Coe, personal communication.

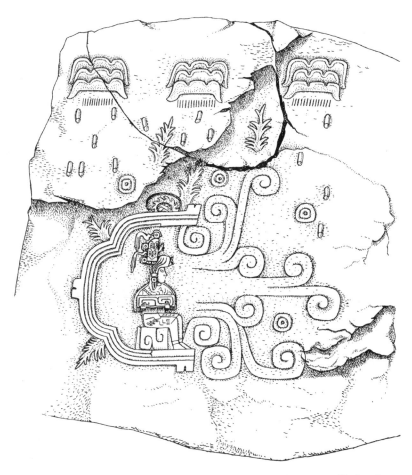

Figure 8.1. The El Rey sculptured relief on the wall of a cave at Chalcatzingo. Monument 31, drawn by Michael D. Coe, 1965. Used with permission.

planning and had, by 600 BCE, a pyramid, ball court, and stone bas-relief sculptures of great beauty and importance for Olmec studies.

READING OLMEC HARMONISM

We have seen that ancient harmony (in McClain's system reconstructed from Plato) was theoretically based on dividing the whole by the first three prime numbers, 2, 3, and 5. A *harmonic number* is an integer made up of only these three primes, as factors within ratios, the number 1 being implicit in all integer numbers in making them whole. The presence of 2, 3, and 5 within a harmonic number makes it creative in being able to host, as the *limiting number* of a given tonal population, a rich set of intervals between that number and its half, which is the tone an octave lower.

The Number 2

Any even number can be divided by 2 to create an *octave,* the simplest and most harmonious interval. A second function of 2 is to develop the continuum of possible harmonic numbers, throughout the discrete harmonic continuum of *possible* octave limits.

But the number 2 cannot create tones in its own octave, for doubling just creates more octave intervals, which renders the process of doubling "barren" and unable to produce any other intervals. To be populated with tones, an octave must be "penetrated" by larger primes already present in the octave's limiting number, making numbers and reciprocals—that is, their interval ratios—that are smaller than 2. For example, $3/2 = 1.666$ (prime 3) and $5/4 = 1.25$ (prime 5) are both less than 2 and greater than 1.

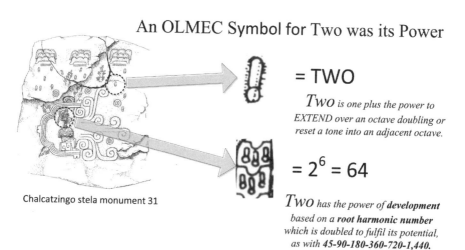

An OLMEC Symbol for Two was its Power

= TWO

Two is one plus the power to EXTEND over an octave doubling or reset a tone into an adjacent octave.

$= 2^6 = 64$

Two has the power of **development** based on a **root harmonic number** which is doubled to fulfil its potential, as with **45-90-180-360-720-1,440.**

Chalcatzingo stela monument 31

Figure 8.2. How the Chalcatzingo monument no. 31 appears to (*above right*) signify the number 2 in its power to stretch out and define an octave, and (*below right*) then be used as a counter for powers of 2 within a harmonic limit.

The Olmec symbolized 2 as an exclamation mark, the dash above a dot showing the extension of an octave. These symbols are normally interpreted as rain coming from the three clouds (with 13, 15, and 14 etched below) so that the rain god is being invoked by the person in the cave.

The Number 3

The Olmec symbolized the number 3 as a quetzal bird. The map inset within figure 8.3 shows the quetzal or "feather" bird's territory, regionally specific to the Olmec, who deified it as Quetzalcoatl, the compound Feathered Serpent who represents a cycle of fifths in serpentine tuning order but harmonically flying. The triple feathers signify the prime number 3 generating successive fifths, in ratios balanced by prime number 2. Two such birds means two powers of 3, or $3 \times 3 = 9$. Each bird is accompanied with the generative symbol of a serpent on a stone with a small circle within.

An OLMEC Symbol for Three was a Bird

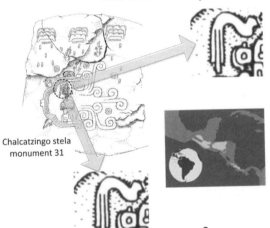

= THREE

Three is a prime number, denoted by a circle within a circle (= prime) with serpentine "hat" and a land bird with three tail feathers.

The resplendent quetzal is an aptly named bird that many consider among the world's most beautiful. These vibrantly colored animals live in the mountainous, tropical forests of Central America, where they eat fruit, insects, lizards, and other small creatures.

Chalcatzingo stela monument 31

$= 3^2 = 9$

Figure 8.3. The Olmec appear to have signified each factor of three with a quetzal feathered bird shown with three feathers. This bird gave its name to Quetzalcoatl and is native to the original Olmec heartland.

The Number 5

In numerical tuning, only one factor of 5 is required to achieve practical just intonation, exactly as is found in the Olmec monument, where the "numerical figure" emerging from the harmonist's head has a single head equal to five since the harmonic number 2,880 is equal to $2^6 \times 3^2 \times 5$.

In the Chalcatzingo iconography, 5 appears as the head of a human-like figure, in a garment showing the six powers of 2, as a livery.

The limit 2,880, a doubling of Adam's greatest extent of 1,440, adds a significant part of the astronomical story, the eclipse year standing in the ratio of a minor diesis above the lunar year. This limit completes the area of wetted bricks (symmetrical tones) to allow all seven possible modal scales to be realized, within just intonation (as per chapter 6). This might link to the Chicomoztoc of seven wombs (figure P3.1, page 144), from which the Nahuatl-speaking people arose.

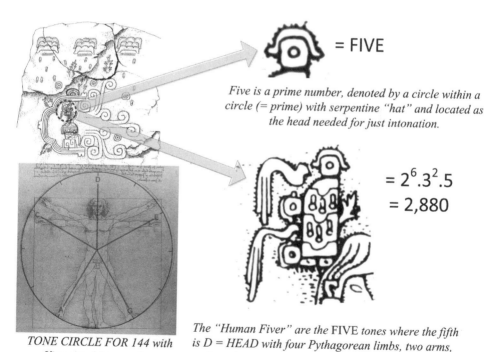

= FIVE

Five is a prime number, denoted by a circle within a circle (= prime) with serpentine "hat" and located as the head needed for just intonation.

$= 2^6.3^2.5$
$= 2,880$

TONE CIRCLE FOR 144 with Vitruvian Man overlay.

The "Human Fiver" are the FIVE *tones where the fifth is D = HEAD with four Pythagorean limbs, two arms, and two legs*

Figure 8.4. The symbolism of the Olmec for 5 within harmonic number 2,880. Note the serpent raised up by a head resembling how part of Tiamat was held above the part used for the just diatonic scale.

GOING BEYOND ADAM

Having established the symbolism of the La Reina cave as representing an octave, and the harmonist's thought bubble as representing the limiting number 2,880, the scroll-like projections from the cave can now be considered as an Olmec way of presenting the tones in a similar but different way as a tone circle. The art presents number 2's third gift: as well as forming the octave interval and keeping tones within a single octave, the number 2 divides the logarithmic world of the octave, as heard by the ear, into two symmetrical halves. About the axis of symmetry, D is the re-entrant octave pair of D1 and D2, and "opposite" D is the square root of 2, our notes a♭ or g♯, near the geometric mean of the octave or, ideally, the square root of 2 (of octaval doubling) found in equal temperament for those tones. In La Reina, the octave is not a circle but is a convex "cave" region with D1 and D2 below and above (figure 8.5). The harmonist is seeing the symmetrical tritone brought on by 2,880, and an initiate viewing La Reina was seeing that seeing.

OLMEC DISCOVERY OF THE ECLIPSE YEAR

Returning to the matrix for 2,880, as portrayed in the Olmec sculpture (figure 8.6, *left*), we find that halving the size of the unit employed doubles the limiting number of Adam = 1,440 to 2,880. Some new bricks emerge, the g♯ brick valued at 2,025 and another generating the mountain's twin peak, shown in the relief as a circle atop the octave cave's horseshoe shape (see figure 8.6, *top right*).

The La Reina monument marked the key arrival, with 2,880, of the eclipse year as a new integer "string length" one minor diesis (three major thirds or 125:128) above the lunar year. We will similarly find the synod of Mars, in Teotihuacan, is a minor diesis above the synod of Jupiter. The monument must have been made as part of a celebration of past discoveries, importantly recording cultural knowledge within an oral culture. Our caption could read "a harmonist gazing out from an octave on her discovery, the harmonic number for the eclipse year."

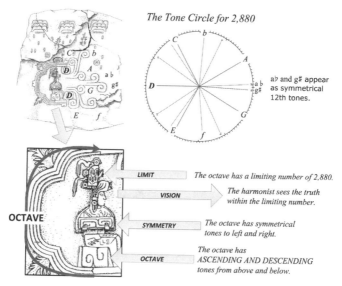

The Tone Circle for 2,880

a♭ and g♯ appear as symmetrical 12th tones.

LIMIT — The octave has a limiting number of 2,880.

VISION — The harmonist sees the truth within the limiting number.

OCTAVE

SYMMETRY — The octave has symmetrical tones to left and right.

OCTAVE — The octave has ASCENDING AND DESCENDING tones from above and below.

Figure 8.5. Picture of an ancient female harmonist realizing the matrix for 2,880. If we tilt our tone circle so that the harmonist is D and the cave is the octave, then the octave is an arc from bottom to top, the limit. Above and below form two tetrachords to A and G, separated by a middle tritone pair, a♭/g♯.

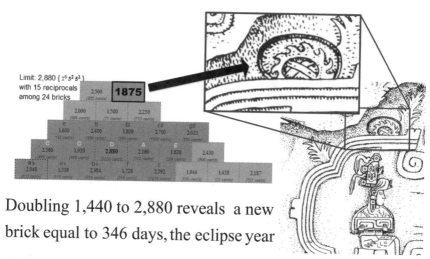

Limit: 2,880 $\{2^6 3^2 5^1\}$ with 15 reciprocals among 24 bricks

1875

Doubling 1,440 to 2,880 reveals a new brick equal to 346 days, the eclipse year

Figure 8.6. In the La Reina carving, a sphere with crossed bands is seen emerging from the octave for 2,880, with twelve flames, while the number 1,875 corresponds with the eclipse year of 346.62 days. A number like 1,875 must appear where shown because its prime formula is $3^1 \times 5^4 = 1,875$. It must be on the fifth row because of prime 5 and in column 2 because of prime 3 and, having just appeared, it is a "rogue male" number* having no powers of 2 with which to form an octave.

---

*Rogue males are a Platonic idiom since two is the female part that, by definition, they lack.

THE OLMEC CITY
OF TEOTIHUACAN

Some centuries after the Chalcatzingo bas-relief, the Olmec city of Teotihuacan became the largest city in the pre-Columbian Americas. It set precedents for, and had a direct influence over, the later Mayan civilization. For this reason perhaps Teotihuacan is termed a preclassical Mayan creation. The city had large stepped pyramids and a long road with "courts." Saburo Sugiyama has worked on the city design for decades, establishing with great care a Teotihuacan measurement unit (or TMU) of 0.83 meters.[2] This length happens to fall just above the unit of measure called the megalithic yard (2.722 feet), and when the smaller astronomical variant of the megalithic yard is used, the center lines of the Pyramid of the Sun and Pyramid of Quetzalcoatl are found to be exactly 1,440 astronomical megalithic yards* apart, along the axis of the road, as if the road were a string length in a harmonic scheme (see figure 8.7). This is an interesting example of how an invariant number (1,440) can improve the likely actual value of a proposed unit of measure. It is also interesting that Professor Sugiyama, like the archaeological community in general, give historical metrology no place when interpreting ancient artifacts.

The design of Teotihuacan appears to follow the 5-by-12 design found elsewhere, for example, in the Parthenon's cella (chapter 4) and the Station Stone rectangle at Stonehenge (chapter 1), a theme through millennia of monument building, each commemorating the second Pythagorean triangle yet offering something the Olmec could appreciate from nature: an ability to (a) count in matrix units 1:80 of a month long, along the 12 side to fill it with 960 units, and (b) count in solar days along the 13 side of the triangle/diagonal, so as to

---

*The astronomical megalithic yard is 19/7 feet long and, when counting time in lunar months, it generates a single foot of difference between the solar and lunar year count. The original megalithic yard was the number of day-inches of difference between the day-inch counts of three solar and lunar years (chapter 1).

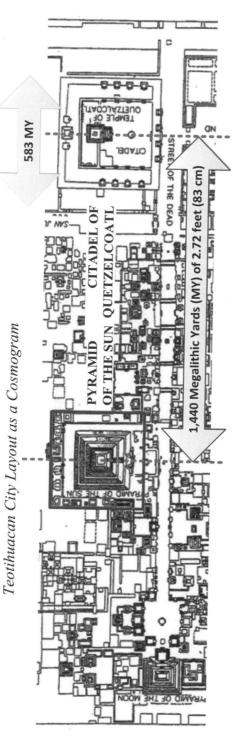

Teotihuacan City Layout as a Cosmogram

PYRAMID CITADEL OF
OF THE SUN QUETZELCOATL

583 MY

1,440 Megalithic Yards (MY) of 2.72 feet (83 cm)

Figure 8.7. Presence of 1,440 units of length between the Pyramids of the Sun and Quetzalcoatl, when corrected to having been the Old World unit of the Assyrian yard or 90 shu-si (see chapter 2 concerning Marduk's ziggurat). From the perspective of metrology, the TMU found by Sugiyama (0.83 meters) can be "corrected" if his measurement of 1,439.75 TMUs was an intended 1,440 TMU of 82.9856 cm.

synchronize with the Feathered Serpent, who includes a set of four time periods all divisible by a thirteen-day week. The four "thirteen-day" periods are Mercury (116), eclipse year (346), tzolkin (260), and Mars synod (780), and these are cleverly integrated in the fundamentals of Teotihuacan's Sacred Precinct. The fifth period in this feathered serpent is Quetzalcoatl as Venus synod, whose day count (584) does not divide well by 13 but whose relationship to the *haab* of 365 days (8:5) is very accurate, allowing Venus to be represented, if uncounted, using the 5 side of the triangle/rectangle. It is interesting that from the Aztec point of view the city of Teotihuacan was where the gods had sacrificed themselves to create time.

Since the 1980s careful archaeological and architectural work has revealed a main timeline for the important design and development of the Sacred Precinct and its four main elements. We must use the accepted nomenclature for the main structures: the Sun pyramid, the Moon pyramid, the Road of the Dead, and Quetzalcoatl's citadel and pyramid—though these names have "stuck" without people knowing what these structures actually meant. The site runs along a significant road of hardstand called the Road of the Dead, whose orientation is 15.5 degrees east of north. The Sun pyramid appears to have been centrally located between the Moon pyramid and the San Juan River, standing east of the road, as if serving as the road's fulcrum. In Sugiyama's architectural units, the farthest extent of the Moon pyramid is 1,000 TMUs north of the Sun pyramid, at the end of the road, while the river is 1,000 TMUs south, running under the road, which then runs on to the citadel, within which, east of the road, the Pyramid of Quetzalcoatl stands.

By 200–250 CE, the site was defined to have 1,440 TMUs between the center of the Sun pyramid and the center of the Quetzalcoatl pyramid within the citadel. The size of the Sun pyramid at that date has been found to have been 260 TMUs square, standing centrally between the road and an extreme boundary for the Sacred Precinct of 520 TMUs, or twice its width. Sugiyama, whose work we cite here, interprets 260 as significant given the sacred calendar of 260 days innovated by the Olmec.

GEOMETRICAL DERIVATION OF
THE TEOTIHUACAN MEASUREMENT UNIT

Though it is possible to allocate linear measures from ancient metrology to "demonstrate their use" at Teotihuacan, the origins of metrology were geometric, not linear, just as harmonic calculations were first geometric rather than calculated. The Teotihuacan builders show clear links to the most ancient traditions of the megalithic, which are best expressed in geometrical solutions to problems, especially as seen at Le Manio quadrilateral in Brittany.

The original counting geometries used inches to count days, and the key breakthrough was found over three years, where the excess of the solar year over the lunar was the megalithic yard of 32.625 day-inches. It soon became the norm to substitute the count per month in day-inches with the megalithic yard, initiating later metrology since, over 12 months of a single year, the excess of the solar year over the lunar year became one foot—the root unit of all subsequent rational measures. The foot relates to the megalithic yard as 2.715 feet, the astronomical megalithic yard, over the nineteen-year Metonic period, but it is what happens at the ends of the three- and one-year counts that shows how the Teoti's derived their TMU to be 2.722 feet so as to compensate for the 2.702 hidden as the reciprocal of the matrix unit 0.368 days (1/80 of a lunar month).

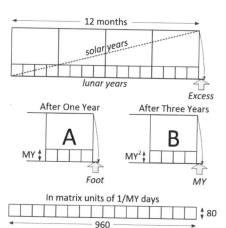

Figure 8.8. If megalithic yard (MY) = 80 matrix units, then (because the matrix unit is 1/MY) it can be equated to MY, canceling to 1 (see A). This makes one day equal to MY^2, and at Teotihuacan, 1,194.99 m. = 3,920.57 ft. = 523.55 MY^2, where $18 \times 29.53 = 531.55$ days and 3,920.57 ft. also equals 1,440 TMU.

From the perspective of the planetary matrix, 1,440 units along a road between pyramids is a harmonic reference to the limiting number 1,440 defined by Adam's name in the Bible, equal then to a period of eighteen lunar months, or 531.55 days, of the "Lunar Series hieroglyphs of the Maya Classic period inscriptions."[3] The Feathered Serpent harmonically flies 125/128 above this to give an ideal of 519 days, just one day short of the adopted tzolkin period of 260 days × 2. For reasons that will become obvious, the Feathered Serpent is rationally organized, according to the 13-day week of Olmec and Mayan chronology, with respect to its first four time periods: Mercury synod, eclipse year, tzolkin, and Mars synod.

The Sun pyramid appears to be the center of Teotihuacan. When traveling south along the road, from the central axis of the Sun pyramid, one finds a suitable coding system for the limiting number 1,440 where the lunar year occurs just before the river (lunar year is 960 TMUs), Saturn synod just after (a♭: 1,024 TMUs), and Jupiter synod after that (1,080 TMUs). The center of the Sun pyramid therefore represents an origin for harmonic numbers, like the bridge of a guitar, except that pitch goes up with length being time, measured using the number of matrix units within celestial time periods. These matrix units of time are 0.369 days, while in feet they are the TMU of 2.722 feet, and these two numbers are almost exact reciprocals of each other so that, multiplied together the counting is one day per TMU squared—where the TMU is a megalithic yard (or MY—see figure 8.8).

But there is another type of counting at Teotihuacan, and that is of days using the TMU (figure 8.9). The Sun pyramid in the earlier period had 260 TMUs per side, meaning days and not matrix units, and this is the coding with respect to the northern direction, on the road to the Moon pyramid. The tzolkin heart of the Feathered Serpent introduces the unique metric of twenty 13-day weeks, a week that further divides into other astronomical periods: the synod of Mars as 780 days (60 weeks), the eclipse year as 346.72 days (26.666 weeks), and Mercury synod of 115.88 days (~9 weeks). Harmonically, Mercury, eclipse, tzolkin, and Mars periods form a serpentine cycle of fifths, that is, of 3/2 or 3/4 ratios relative to one another, flying in the sense of raised by a minor diesis relative to the World Soul (9/8) between the lunar year and Jupiter synod. The eclipse period is above the lunar year and the Mars synod above Jupiter, corresponding with Karthikaya in India, as Mars is the son of Brihaspati, or Jupiter. This reveals the Feathered Serpent as a structure paralleling the World Soul.

To understand the Feathered Serpent one has to see the harmonic mountain as composed of two registers, each of three rows by five columns (see figure 9.2). (This sectorization of multiple rows was encountered in chapter 7, with Osiris and his four-row pediment.) The lower register is on the "earth," familiar to practical music making, and the higher is in the "sky," a familiar but veiled form within religious texts, monuments, and iconography.* We now have three meanings of the world *heaven:* the astronomical world of planetary gods, the harmonic world created by prime 5, and the abstract afterlife space created by monotheistic religions.

The first register has the Saturn synod as its cornerstone unit of 0.3692 days, this "stolen" from Uranus's synod of 369.66 days, divided by 1,000. Seen on the limit 1,440, Uranus has 1,000 units, standing 25/24 "above" the lunar year and 125/128 "above" Saturn. This may be why there are 1,000 TMUs to the San Juan River from the Sun pyramid. By 350 CE, there were also 1,000 TMUs to the far side of the Moon pyramid, after its expansion. It is Uranus that is the cornerstone of the second register of the Feathered Serpent.[†]

The harmonic difference between Saturn and Uranus gives the reason why the TMU works in two ways; as a harmonic lunar unit (1/80 of a lunar month) and as a 13-day count for the periods of the Feathered Serpent's tetrad of component periodicities: of Mercury, eclipse, tzolkin,

---

*There is a third cosmic register possible, based on Marduk, Osiris, and Indra and the seventh power of 5, which can generate a near perfect square root of 2 to kill the dragon/serpent on the earth, liberating the human prime 5 (see chapter 10 for where this may lead).

†While Uranus is theoretically visible to the naked eye, one needs to know where to look for it. Its synod is 25/24 lunar years or 12.5 lunar months long.

and Mars. Saturn's cornerstone unit = 1 is 0.369 days = 7/19 of a day. Nineteen such periods gives us the more familiar 7-day week, which divides into Saturn's synod (378 days) as 54 weeks, into the Saturnian year (364 days) as 52 weeks (see chapter 10), and Jupiter's synod (399 days) as 57 weeks. The cornerstone makes the 7-day week work within its register just as Uranus, cornerstone of the Feathered Serpent, enables a 13-day rationality.

The Uranus cornerstone (of the upper, "feathered" register) must be 125/128 smaller than Saturn's, and is 13/36 of it. In thirty-six such units, the 13-day week arises and, sure enough, the synods of Mercury and Mars, the tzolkin, and eclipse year divide by 13-day weeks—not perfectly, but well enough for the harmonic reality to also follow a 13-week calendar. While the Venus synod (584 days), as Quetzalcoatl, is 3/4 of the Mars synod (780 days), the result is forty-five 13-day weeks while Mars is sixty 13-day weeks.

The northern side (to the left in figure 8.9) of the Teotihuacan sacred complex, between the Sun and Moon pyramids, appears devoted to the Feathered Serpent's tetrad. South of the Sun pyramid is the harmonic model for 1,440, and its original stairway could well have been 45 TMUs wide; the harmonic root of Adam = 45. After 1,440 TMUs, the center point for the Quetzalcoatl pyramid is reached, and the surrounding complex there appears extensive by 200 CE.[4]

The size of the Road of the Dead would have allowed a large number of yards to be deployed so as to count and rescale either geometrically *or* harmonically, within the Sacred Precinct. One harmonic technique would have been to easily double the limit of 1,440 TMUs by using half yards, placing 2,880 units between the pyramids. Another example, using geometry, reveals a powerful triangular form connecting the periods of the Feathered Serpent; then leading to the 2,880 limit seen at Chalcatzingo's La Reina/Monument 31.

Sugiyama suggests the distance from the far edge of the citadel was 520 TMUs, so then the Mars synod, at 3/2 of this (780 days) corresponds with the island platform between the two mini-pyramids "guarding" the entrance to the Moon pyramid, whose center is 780 TMUs from the Sun pyramid datum on the Road. At right angles, these two

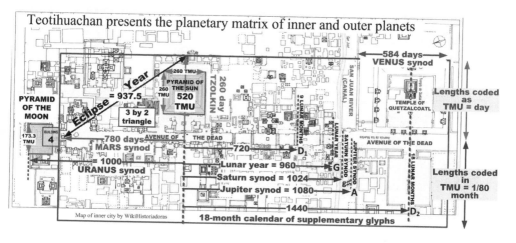

Figure 8.9. Division between geometrical modeling (in 13-day periods), *left*, of Sun pyramid and harmonic modeling; then, *to its right*, using matrix units of megalithic yards.

periodicities, of Mars and double tzolkin, form a right triangle with sides 3 tzolkin by 2 tzolkin. The third side (the hypotenuse) is then 937.5 TMUs, which, if recorded as a rope and moved to the datum beginning the southern portion of the road days, marks a harmonic matrix point 125/128 less than the lunar year, 960 TMUs long. To "clear" the fractional half in 937.5, the harmonists would need to double the limit from 1,440 to 2,880, counting 937.5 in half TMUs and so doubling 937.5 to 1,875. This value only emerges when Adam's limit of 1,440 is doubled to cause, as at El Reina, the limit 2,880 within which the eclipse year appears.

The hypotenuse of this triangle is uniquely the square root of 13, and the Feathered Serpent periodicities are divisible by 13, so that the sides 2 and 3 when squared become 4 and 9, which, added together, make 13 "the square of the hypotenuse." The 13-day unit in both is squared and multiplied by (4 + 9 = 13), and is therefore cubed. The hypotenuse, as the square root of 13, cancels the square to give 13 × 20 × 20. This means that the hypotenuse of this triangle will be the square root of (400 × 13) × 13 = 937.443331620637.

But relative to 520, the eclipse year should be 520 × 2/3 = 346.6667, that is, $(40 \times 2/3) \times 13 = 346.6667$. The matrix unit has somehow become inserted between them. It is

$$346.6667/937.443331620637 = 0.369800166375$$

This matrix unit equals $4/3 \times 1/\sqrt{13} = 0.369800130816819$, showing that the effect of having two later parts of the Feathered Serpent as the 2 and 3 sides gives their preceding part since, in the cycle of fifths, backward is a cycle of fourths.

But this triangle would not have been thought of as an algebraic solution to the problems of harmonics on the Feathered Serpent. Instead it was probably found to be true that, when forming a triangle out of the day counts of the tzolkin and Martian synod, the number appears that gives the bridging relationship for the 1,440:2,880 matrix. The 3-by-2 triangle has one further application of this available to it, since Tlaloc or Mercury lies at the tail of the Feathered Serpent. The double tzolkin period equals three eclipse half years, the half year being the smallest period between eclipses occurring at opposite lunar nodes or "crossings" of the sun's path by the moon. If what was the 2 side becomes the 3 side, then one eclipse year north, along the road north, generates a hypotenuse of length 624.84 TMUs, which (as a rope length) can be laid south from the Sun pyramid. If the south end of the rope is fixed and the other end laid to the south, the rope is doubled to 1,249.68, on the 1,440 matrix as 1,250 TMUs, locating Tlaloc on the southern road, close to 1,440, which is where Quetzalcoatl's pyramid signifies Venus.

With great subtlety, the whole of the planetary matrix was realized by the Olmec in their city of Teotihuacan. The 18-month limiting octave of 1,440 TMUs was attached as their supplementary series of glyphs, often found on Mayan Long Counts.

QUETZALCOATL'S CALLING CARD

After a millennium of development, the Olmec had innovated using megalithic yards to model both matrix units and days, to *geometrically* model their Feathered Serpent calendar at Teotihuacan. Their origin

Figure 8.10. Olmec bas-relief at La Venta, Monument 13, sometimes dubbed "the Ambassador" or "the Walker." Figures 8.11 and 8.12 are associated. Courtesy Linda Schele Collection (SP-127019) © David Schele (FAMSI.org).

myth, now long ago, was of receiving sacred knowledge from an individual person called Quetzalcoatl, a bearded Semitic wise man from the East (figure 8.10), while also associating that name to Venus and also to the whole of the Feathered Serpent that was part of the wisdom of Quetzalcoatl.

In figure 8.10, the man Quetzalcoatl was possibly drawn within a grid based on the human foot shown behind him, as per figure 8.12. The "foot" is tilted at 45 degrees so that the big toe defines the left hand starting point for the grid, then of half-foot squares. The rest of the work then extends 5.5 feet high by 3.5 feet wide, a grid of 11 by 7 in which the figure is spanning a height of 10 squares, 10/7 of the width and therefore representing the tritone a♭.* The axe-like focus is a metrological measuring device whose width is one unit (the grid's "half foot") while its height only becomes an integer when the half-foot grid

---

*10/7 = 1.428, while a♭ = 64/45 = 1.422, and $\sqrt{2}$ = 1.414.

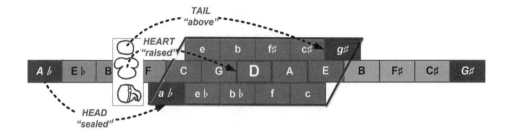

Figure 8.11. How the three symbols on La Venta Monument 13 represent the cutting up of Pythagorean tonality into just intonation.

is divided into 7 units, the height then becoming 24 units. The circle below is one grid unit in diameter as is the lower petal of the middle trefoil while the decapitated bird's head has a diameter root 2 times the grid size.

The bird's head is sealed by the √2 tritone (a♭), from which we can deduce that the bird is the defeated Pythagorean monster whose head was A♭ but is now a♭ (see figure 8.11). The central symbol is the heart of the monster, D, now raised up above its head while the tail forms a sky of chromatic alternatives including g#, the twin tritone. This then recapitulates the historic generation of just intonation, amenable to numerical tuning theory and the myths of flood heroes. But Quetzalcoatl stares intently at the blade he holds aloft, which resembles the daggers employed by Shamash (also Ea and other matrix gods) to cut through "the earth" to rise up within the matrix. The three symbols below span a height of three rows (25/24, the chromatic semitone) since two powers of 5 increase and one power of 3 is lost between, for example, f and f#; 25/3 becomes 25/24 by applying 2 three times in order to the bring the tone number back into the octave.

Quetzalcoatl's blade is actually a representation of chromatic semitone 25/24, now as a triangle, and applied above the triple rows below in order to reach the height of the Feathered Serpent, $5^4 = 625 = 25 \times 25$. His grip on the blade is the row $5^3 = 125$, the row of Uranus, who functions as a♭ for the Feathered Serpent, and thus is its savior. And g# is the other tritone, the 365-day year or haab on the level of the tip of

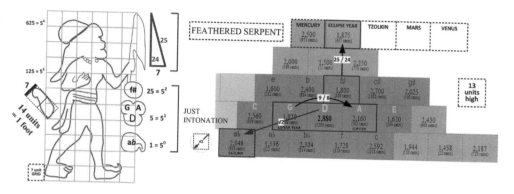

Figure 8.12. The orderly grid behind Monument 13 at La Venta, thought to be Quetzalcoatl. Three rows of the harmonic "earth" are shown as icons, while above, his "blade" appears metrological, in harmony with the grid. It reveals the chromatic semitone found between two rows, which he places in "heaven," to see the "Feathered Serpent."

the triangular blade, $5^5 = 3,125$, and of YHWH. (Note that the figure holds the blade with his left hand as he is looking at far right corner of the grid, into harmonic space.)

THE KEY FEATURE OF OLMEC ART

The Olmec evolved the Long Count for which the Maya are famous, a counting system of days that concluded one of its major cycles, that of 13 *baktun* of 144,000 days each, in 2012. Olmec Long Count dates can be read using the number notation, in common with the later Mayan civilization, of line = 5, dot = 1, and a vertical ordering of numbers according to larger units of time in a base of 20. The date 68 BCE* is carved on the rear of Stela C at Tres Zapotes (see figure 8.13), making this the oldest Long Count that can still be interpreted and classified as in the Olmec's late-formative period.† What is on the front of Stela C is

---

*The consensus is 68 BCE, though it could be 450 BCE if the "zero date" for the count was changed during the evolution of the Long Count system.
†While being pre-formative Mayan in the sense of accepted Mayan nomenclature, this long count is prefixed as the Maya would have, with an "introductory glyph."

also interesting, relating, I believe, to Adam's harmonic root, the resolution of eclipses (at limit 2,880) as relating to the √2, the role of stepped pyramids, jaguar altars, and human sacrifice.

Two pillars frame a stylized face, with eyes, nose, and "jaguar" mouth below. Three "tears"* adorn each cheek, and the accentuated brow and forehead move up to a rectangular pendant tied to the pillars, within which is a diagonal cross, at the center of which are nested rectangles and an innermost triple rectangle. Atop the pendant is a gaping-mouthed human profile with hat and nested circle behind, below a √2 right triangle.

When the height of the twin pillars is compared to the distance between them, the ratio found is the √2, and this is a known geometrical

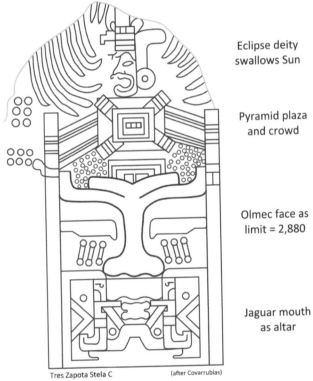

Eclipse deity
swallows Sun

Pyramid plaza
and crowd

Olmec face as
limit = 2,880

Jaguar mouth
as altar

Tres Zapota Stela C   (after Covarrubias)

Figure 8.13. Tres Zapotes Stela C (redrawn from Covarrubias) has four submontages within a double square, interpreted as noted on the right.

*Similar to the "rain" at Chalcatzingo.

construction found in later Mayan art, which resembles the Egyptian practice of using ropes to generate the "root rectangles" with diagonals of √2, √3, √4, √5, and from these phi, the golden mean, all from the simplest of rectilinear forms, the square with a side of length 1.[5] The length of the original rope used for the construction of the square defined the 1 unit.

If a square is drawn at the bottom of Stela C, between the towers, then it embraces the brow line of the stylized face and the diagonal of a square is the square root of 2. If both the diagonals are arced until upright, they touch the tops of the twin pillars, and where they cross passes through the central square (of 3), in the middle of the rectangular pendant held between the pillars. The diagonal crossing of the pendant is therefore indicative of the square root of 2, harmonically the tritone

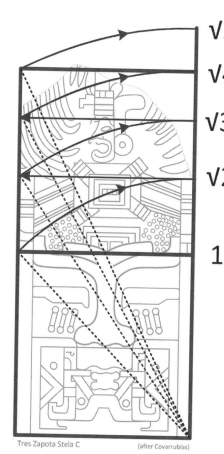

√5

√4

√3

√2

1

Figure 8.14. Traditional triangular method for evolving the square roots of the numbers 2, 3, 4, and 5, constructed using ropes.

Tres Zapota Stela C   (after Covarrubias)

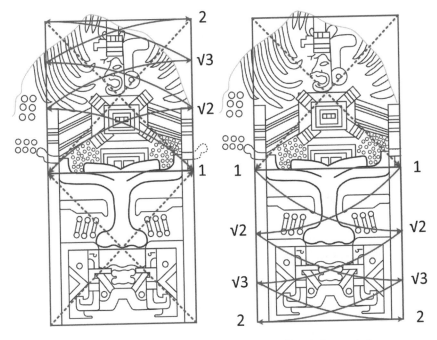

Figure 8.15. Stela C with evolved square roots from below and above.

to D, but what is the limit? The stylized face gives the limit as having the harmonic root using the following counters: the two eyes equal two powers of prime 3, and the nose equals a single power of prime 5. This gives us the harmonic root of $3 \times 3 \times 5 = 45$, the same as Adam and therefore probably having similar harmonic connotation as being the generic man and the eighteen-month supplemental period in the harmonic terms already established. Below the eyes on each cheek, three pairs of double circles are shown (according to the epigrapher's drawing), each pair linked like tears, representing six powers of 2: $2^6 = 64$, and $64 \times 45 = 2,880$, the same limit as Chalcatzingo's La Reina (mentioned on page 182) the 2s drawn similarly, as exclamation marks but with dots at both ends.

If a square is drawn on top of the first to form a double square (compare the Egyptian djed pillar) then the world above includes the "pendant" and the gaping profile. Repeating the arc downward causes the root-2 diagonals to cross at the nose, defining a level at the lowest point (accentuated) of the two eyes. The pendant at the higher root-2 crossing is therefore symmetrical to the lower root-2 crossing at the nose of the limit.

There are now two rectangles not involved in the root-2 crossings, and the diagonals of each can show the central detail as (above) the gaping profile and as (below) the blobby decapitated symbol that in Olmec art is the human heart and its four aorta after removal from the chest. Thus, the gaping profile is eating the human heart that is held within the stylized altar, modeled after the jaguar's mouth, a distinguishing motif for Olmec art along with the gaping mouth. The crossing point above is a pyramid complex *seen from above* in which the sacrifice is taking place on Earth, a triple-stepped pyramid with four diagonal staircases shown. The drawing shows many small circles, like groups of people (from above), and above the face another pyramid with a top split by a channel into two and three parts.

The two circles around the altar below probably represent priests seen from above, and the offering's body is shown in profile, with the heart beneath the chest. Thus, (1) on the top rectangle is the eclipse deity; (2) below is the Sacred Precinct for human sacrifice, at that moment seen from above; and (3) below that is the limiting number that is symbolic of the First Man and revealing the eclipse year as part of the Feathered Serpent, connected to the symmetrical tritones of a♭ and g#, seen in the world of counters for harmonic powers. Finally, (4) at the bottom is the human sacrifice being attended to, as the gaping mouth of the jaguar who eats the moon.

In summary, a four-part graphic was made on Stela C that is similar to the general style and format of Olmec art relating to the jaguar mouth and the metamorphosis from human to jaguar features. A complex of religious beliefs and activities on the theme of the human relationship to the eclipse is compressed into four seamless panels, a super-symbol based on the square root of 2, connecting the above with the below and enabling what was probably the central meaning of the Olmec culture, stated on this single stone carrying a Long Count date of 38 BCE.

Christopher Pool writes "According to Justeson, Stela C appears to commemorate a lunar eclipse, which was followed two weeks later by a solar eclipse that was nearly total at Tres Zapotes."[6] The profile head appears to have eaten the sun (or moon) that is then behind the profile, beneath what Pool identifies as a rearward-facing tenon typically employed to hold such heads, when objects, in place within a built structure.

This interpretation would have the jaguar mouth and jaguar face naturally linked, through the harmonic number 2,880 and its root 45, to some sort of human sacrifice at eclipse for which purpose one needed to know when an eclipse was going to happen using a day-count calendar. The 260-day tzolkin has a natural relationship of being 3/2 of the eclipse season, making the eclipse year (a double season) 3/4 of a double tzolkin and, similarly, the pyramid plan view at the root-2 crossing is a rectangle expressing the ratio 3/4. The right pillar could be the tzolkin and the left pillar the synod of Mercury so that harmonically the unit distance between pillars is the tone 9/8.

The 260-day tzolkin period has been deduced, from found artifacts containing it and their distribution, as "in use much earlier than the 'Long Count.'"[7] Pool suggests that "the chronological progression from Stela A, to Stela D, to Stela C at Tres Zapotes graphically reflects the continuation of the trend toward increasing historicism in Formative monuments from Olman, culminating in dated Inscriptions,"[8] such that the calendar becomes a way of recording history, while our own evidence points to the necessity for calendric prediction of events so that rituals can be synchronized with the heavens. These two points of view, celestial prediction and cultural historicity, can equally explain the emergence of the Olmec day-counting system, first the 260-day and then the Long Count, into a system not found in the Old World.

The geometry within Olmec art is additional evidence that sacred number sciences from the Old World had reached the pre-Columbian world: sacred geometry and arithmetic, metrology, harmonic limits, and artificial mountains. The cities of San Lorenzo and La Venta suggest a date around 1200 BCE, but there might have been a number of contacts made from then on, from both East and West.

EVOLVING THE NOTION OF SACRIFICE

Harmony was fated (*a*) to be a system involving compromise between the effects of prime 3 and prime 5 within the octave defined by prime 2, and (*b*) in modern music making, to result in the abandonment of pure tonal intervals within an equal temperament where every semitone is a twelfth root of 2. Equal temperament is the greater compromise but makes for a much more versatile tuning system. But the synodic planetary world had to be established by fixed rational intervals that stay the same and cannot be "updated," because they rely on the numerical interaction between planetary periods which must be near rational.

If genuine spiritual issues for existence are to be located within tuning theory, then these must hinge on the compromises that make just intonation the most harmonious blending of the three harmonic primes. The octave, once populated, bears the sins of whatever disharmony remains within it—the point of Plato's allusions to different forms of city governance, with their otherwise preposterous proposals for laws and regulation. Avoiding the Pythagorean comma and arriving at the heptatonic scale created just seven notes tuned in one key/scale, which was neither musically useful nor corresponded to the planetary matrix. In killing Tiamat or Vrtra, a just-tuned diatonic was created in which numerous modes were available and, upon reaching the eclipse limit of 2,880, all twelve note classes are present including the dissimilar A♭ and G#, now reborn, in the right order, opening up all seven modes. These seven modes were at the heart of music in the ancient Near East but were lost to the Christian world until the church modes after 900 CE, even while they had continued on in India.

The synchronous appearance of seven modes with the eclipse phenomenon could be associated with the Olmec myth of origin from seven caves (see figure P3.1), and the caves could be the scales that, in Old World Greece and India, were often alluded to as seven tribes (or rivers). We have already seen that the invariance of seven symmetrical notes within an octave is echoed on a higher level by seven symmetrical scales, each linked by the methods found in Sumerian harp-tuning texts.

The emergence of the tritones to D in two extra modes exactly when the eclipse phenomenon becomes rational at 2,880 seems to have been perceived as requiring a necessary sacrifice, corresponding in some way with the action of the tritone. In the Vedic tradition, the god of fire, sacrifice, and the altar emerges as part of the creation, as Agni. It therefore seems that the Olmec, and possibly other religious systems that adopted harmonism, took the idea of sacrifice and linked it to the primitive cult of ancestors and death, widespread in the ancient world. Literal sacrifices could have been an early driver since megalithic times if the eclipse was part of rituals connected to the sun and the moon, since the eclipse phenomenon looks like a death, and indeed the jaguar is known for crushing the head of its victim. It is also true that an eclipse occurs at the junctions between the celestial equator, the heavenly counterpart of the Earth, and the ecliptic along which the planets travel, and there are widespread traditions of the dead traveling along such "rivers in the sky" in their journeys after death (and before life) from the oldest times.

If human sacrifice is not required, the Bible presents this in the story of Abraham's near sacrifice of Isaac. Abraham's willingness to sacrifice is considered the essential sacrifice, this pointing toward the truth about sacrifice: human beings need to learn how to sacrifice in order that some as-yet-unmanifested good can arise, something as yet not manifested within the creation.* But Abraham is learning an art and a skill not dominated by calendars, monuments, and humans killing other humans. Native Americans (it was said) adapted easily to Catholic iconography since at its heart lay God sacrificing his son in order to save the world. In the Rg Veda, the sacrifice is held as involving the soma, the food of the gods, through which wisdom is nourished. Perhaps it is the willingness to sacrifice rather than the outer performance that changes the person.

Olmec statues repeatedly present human figures in a state of metamorphosis that appears connected to the eclipse, whereby the person starts to change into the form of the jaguar (nagual). Changing form

---

*The Olmec and later Mesoamericans may have thought that an as-yet-unmanifested good would arise from the gods who created the world.

is essentially a change of identity that, in the world of the octave, is a shifting of D and hence of its limiting number. The shape-shifter is a frequently found motif within shamanism and myths, and the moment of the eclipse is a breakdown of norms in that either the moon is no longer illuminated by the sun or the sun is no longer illuminating the Earth. The sun and the moon are dominant presences above the Earth, and their domination is magically neutralized at the time of the eclipse. The eyes of the maize-god twins are likened to the sun and the moon as with other ancient world deities such as the eyes of Shiva, from whose hair came the great river Ganges.

The Olmec provide the only surviving record of the Feathered Serpent, spawning the greatest sacred calendar ever conceived and yet they appear as survivors of our own Bronze Age, transported to the Mexican jungle. Meanwhile, the Iron Age in the Mediterranean produced, instead of Chalcatzingo, the Bible, in which the personalization of the planetary gods was forbidden, and classical Greece, in which literacy gave birth to philosophy and physical speculations leading to modern science. Instead of exotica, the Olmec were a parallel reality of the Bronze Age never collapsing and as such they left a comprehensive monument to their harmonic gods. It was never going to create the modern world, and the Spaniards saw the degree of their anachronism.

9

YHWH's Matrix of Creation

We have seen that the planets seen from Earth occupy definite niches within a harmonic space, these defined by powers of 3 and 5. The role of the prime 2 is not *locational* within this harmonic space, leaving powers of 2 to define octaves and equilibrate the powers of 3 and 5 to stay within a given octave, that is, to only make tones between 1 and 2 in our fractional space. One is left with "male" powers of 3 and 5 populating "female" octaves, within what Plato could easily have called a patrix,* a patriarchal model of all possible centers of harmonic reality. Within this patrix, defined by the number field itself and emanating from unity, locations are occupied by different planetary time periods. In that sense the lunar year is separated from both the inner and outer planets by integer ratios of harmony, some microtonal. In the present scientific consensus, however, planetary periods should not in any way be harmonically connected, having evolved through natural forces from the sun's own womb, the solar nebula. But the harmonic separation of planets seen from Earth is not the whole story since, as we have shown, these relationships subsist within the fundamental harmonic scheme of

---

*The Pythagorean musicologist Pete Dello has suggested the term *patrix,* since *matrix* is a feminine term in the sense of being a container, while the patrix of composite numbers made up of odd primes is in principle unbounded.

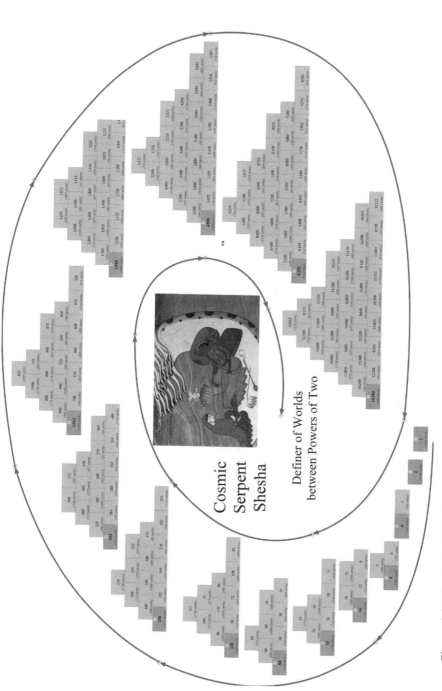

Figure 9.1. The Myth of Adi Shesha fits the powers of 2 well, as these organize the harmonic numbers that are possible limiting numbers for octaves. In the center lies Vishnu, resting between creations, on Ananta-Shesha, with his consort Lakshmi massaging his feet. Victoria and Albert Museum, London.

just intonation, using the lowest numbers possible. This therefore casts doubt on celestial dynamics as having been a dumb mechanism of planetary and moon formation, revealing an aspect of planetary dynamics little understood today but evidently glimpsed by the ancient world.

The inner structural organization of the planetary patrix is of *harmonic roots* separated by powers of 3 to form cycles of fifths and powers of 5 to provide synodic intervals of thirds. In the planetary context these manifest between 5 to the fifth power (as found within YHWH's name code of 60^5) and 3 to the fourth power (as found within the 60^4 needed to locate Venus within that harmonic space). It takes the form of two similar blocks dividing the world into a harmonic heaven and earth, each shaped like a parallelogram, five wide and three high, stacked one upon the other, as shown in figure 9.2.

Venus has been called by many names; Inanna (Sumerian) or Ishtar (Akkadian-Babylonian) were goddesses of love and war, while Aphrodite, our classical Greek version, was more of a love goddess, perhaps due to patriarchal chauvinism. The preclassical Olmec instead had a male Venus, called Quetzalcoatl. Among the Aztecs, Quetzalcoatl was related to gods of the wind, the planet Venus, the dawn, merchants, arts, and crafts. He was also the patron god of the Aztec priesthood,

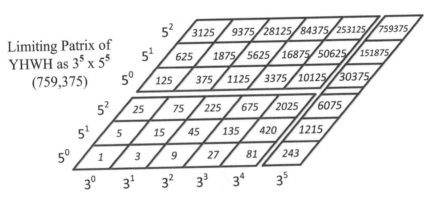

Figure 9.2. Numerically "below" YHWH's limit lies an extensive two-dimensional field of numbers composed of primes of 3 and 5. This field, limited to the fifth power of both primes, encompasses the creative space of YHWH between 60^5 and Saturn, the cornerstone equal to unity, the zeroth power of 2, 3, and 5, and any other prime number. The column for 3^5 is not essentially required by just intonation, which the two parallelograms of "earth" and "heaven" can provide.

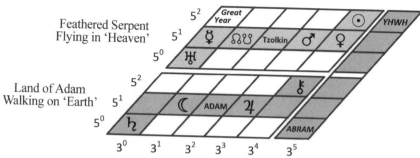

Figure 9.3. The harmonic framework naturally formed two identical substructures then called (*a*) earth, where Adam lives at the heart of synodic relationships of the outer planets to the lunar year, the transpersonal reality called existence, and (*b*) heaven, where inner planets Mercury, Venus, and Mars and lunar eclipses are structurally inharmonious to the earth, being raised above by a minor diesis of 125:128 (just less than an octave).

of learning and knowledge.[1] The name means "feathered serpent," and in this respect Venus leads what the Egyptians instead called a "flying serpent" in the sky.

The idea of sky is strongly coupled with that of earth, since the planets of the inner solar system, Venus and Mercury, form the head and tail of the Feathered Serpent, who flies three powers of 5 above Earth. We have seen that giving human characters a 5 in their patrix number, like Adam $5 \times 9 = 45$, makes them harmonically fecund above what can be expected on the very earth of the first row with no power of 5, which includes only Pythagorean intervals. The two repeated patterns of sky and earth each form a "bed of Ishtar" lozenge, three powers high and five powers across. This bed of Venus (also known as her vulva) gives birth to the seven-tone modal scale (chapter 6) that for the ancient Near East replaced the serpentine cycle of fifths. The Pythagorean serpent crawls only on its belly, on the earth of row 1—yielding a poor seven-tone (heptatonic) scale and a disharmonious twelfth tone (a chromatic Pythagorean comma). Musically inferior to just intonation (the harmonic arrangement found between planetary periods), each row of the matrix is still a Pythagorean serpent but now, above the earth, such serpents can connect to just alternative tones, replacing higher

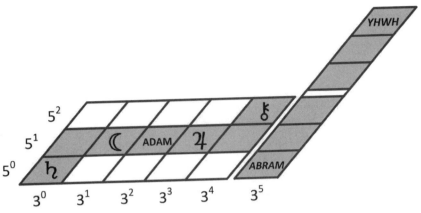

Figure 9.4. The cornerstone region of Adam's earth.

powers of 3 with lower powers of 5. The blend of lower powers of 3 and 5 reduces the size of numbers involved so that (*a*) numerical tuning theory became possible and (*b*) a richer world of modality was created for practical music making.

The cornerstone of harmonic space is the Saturn synod (♄), which has no powers of 3 and 5 relative to the lunar year located on the patrix at 3 × 5 = 15. That makes the Saturn synod a *rising* semitone interval of 16/15 (a♭) relative to the lunar year (G), enabling the post megalithic world to discover this harmonic planetary matrix (see chapter 1). The Jupiter synod (♃) is 9/8 of the lunar year, locating Jupiter at 9 × 15 = 135. As noted in chapter 1, "water" as prime number 2 can be added so as to find the octave in which these three periods naturally sit, namely, Saturn = 128, lunar year = 120, and Jupiter = 135, all under a limit of 180, but then without any ability for modal diatonic scales between D = 90 and D = 180. Doubling thrice gives a limit of 1,440, enabling five modal scales.

The octave 720 to 1,440 is Adam's higher limit, built into his name as 1 + 4 + 40 interpreted in *place notation*. In between the lunar year and Jupiter synod, Adam's limit has a periodicity of eighteen lunar months. Each lunar month is then worth 80 matrix units (12 equaling 960) and 1 more unit, making 81, equals thirty sidereal days (rotations of the Earth). But 81/80 (the syntonic comma) shifts us down one

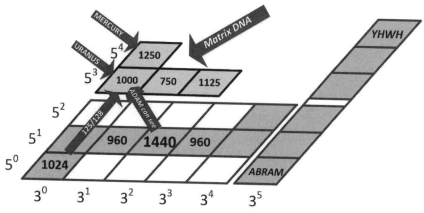

Figure 9.5. The matrix DNA exposed in decimal format above Adam, which Adam "sees," while Abraham "sees" YHWH, sharing the same powers of 3.

and across four places to the harmonic root 243, which is the gematria sum of ABRAM's name. The patriarch Abram, whose name was changed by the Lord God to Abraham, was therefore given the same power of 3 as YHWH, the name only later revealed to Moses. This suggests ABRAM was the harmonic discoverer who first looked to the harmonic limit of YHWH, thus finding the ideal harmonic God for the planetary matrix.

But Adam's matrix for 1,440 is deficient in its ability to form all seven scales, since its area of reciprocity is less than the bed of Ishtar, preventing those two additional modes (Lydian and Locrian), which require reciprocal tritones (a♭ and g#) to give them a special creative role in toppling the tonality of a harmonic system but also then interconnecting, as seen in the Sumerian scale tuning system (see chapters 5 and 6).

Above the three rows of Adam's earth there is a strange number from the start, 1,000 matrix units (see figure 9.5), whose decimal character has gone into a "decimal heaven" where tone numbers directly show intervals as a place notation relative to 1,000, equivalent to our modern representation of ratios, then to three significant places. In this upper region of heaven is a repeat of the transformations found in the lower region of earth, but they are all tonally shifted by the interval 125/128, a minor diesis. This number 1,000 also exposes the matrix unit

(= 29.53059/80 = 0.36913 days) that is the synodic period of Uranus (1,000 × 0.36913 = 369.13, ~ 369.66 days). Thus Uranus is the cornerstone of the heavenly matrix, recovering the role Saturn took from him in establishing the heavens occupied by the inner solar system.[2]

Adam as 1,440 cannot see more than Uranus (the biblical Joseph) and Mercury. To see more requires him to either double his limit to 2,880 or change his location in the underlying patrix. We will see in the following section, "The Heavenly Choir," that the New Testament solution is to raise Adam using two powers of 10, the Revelation of Saint John placing him in this heaven, from there to study "a new heaven and a new earth." We saw the alternative in chapter 8, with Mesoamerican harmonists using 2,880 to expose one more feature of heaven's Feathered Serpent, the eclipse year whose patrix number is $3 \times 5^4 = 1,875$.

The idea of suitable limits was and is essential in exploring harmonic astronomy, and so it reveals a likely timeline for ancient discoveries, especially when usage of such limiting numbers can be found in texts and other artifacts. While the "super-gods" of chapter 2 required very large numbers to explore what subsequently became the understanding of modal scales, the discovery of harmonic connections to the inner planets completed the planetary matrix so as to reveal a full harmonic cosmology. The new "earth" indicates the role of the outer planets with regard to the lunar year and rotation of the Earth, while the new "heaven" is dominated by the inner planets, Mercury, Venus, and Mars. Only the first could yet be seen by Adam, our poster boy for ancient harmonic cosmology and the repurposing (in the Bible) of a worldwide tradition from a common megalithic progenitor.

Adam is not the only launch pad into heaven. Once harmonists understood that the difference between earth and heaven required a rocket carrying $5^3 = 125$, they had already traveled (in their minds) from Saturn to Uranus, the new cornerstone. The lunar year of 960 matrix units, times 125, is 120,000, landing on the eclipse year that 2,880 reveals as 1,875 (and 1,875 × 64 = 120,000). The biblical harmonists incorporate this number into Nineveh, with 120,000 inhabitants who know not their right from their left.[3] Nineveh was used this way at the time of the Twelve Prophets (a set of twelve books approximately contemporaneous with the early Bible writers) to be the location for the linked stories of Job, Jonah, and Nahum.

The Nineveh matrix can resolve a brick into 101,250 matrix units, which is twice that required to see the following:

1. The Venus synod is then 583.98 days, rather than its actual synod of 583.92. The ancients generally used 584 days.
2. The Mars synod is revealed as 778.64 days rather than an actual span of 779.94 days, 780 being symbolic.
3. The period above Adam, as 259.55, rather than the tzolkin of 260 days.
4. The harmonic eclipse year at 346.06 differs from the actual eclipse year of 346.62 days.
5. Mercury is harmonically resolved as 115.35 days rather than 115.88 days.

Such acts of harmonic astronomy do not altogether solve the problem of being Adam but rather set the stage for religiosity, placing Adam as central on the earthly parallelogram and below the tzolkin, a New World sacred year central to the heavenly parallelogram.

CONNECTING HEAVEN AND EARTH

It is obvious that the heavenly serpent requires a new kind of limiting god to contain it and lead like the super-gods of Marduk, Indra, and Osiris did, and under whose jurisdiction heaven and earth were all connected in a religious sense (*religion* literally means to "re-connect"). Thanks to Abram, YHWH appears as that super-god, with his canvas of $3^5 = 243 \times 5^5 = 3,125 \times 2^{10} = 777,600,000$. The number 1,024 equals Adam's cornerstone for octave limit 1,440, the cornerstone having the unique harmonic root (patrix number) of $3^0 \times 5^0 = 1$, and 1,024 exceeding Plato's tyrant number 729. The number 1 stands at the very start of the leading diagonal of YHWH's canvas, while YHWH defines its greatest, balanced extent. Inboard of these two extremes of 1 and $60^5 = 777,600,000$ are the lunar year ($60^1 = 60$) and Venus ($60^4 = 12,960,000$), respectively.

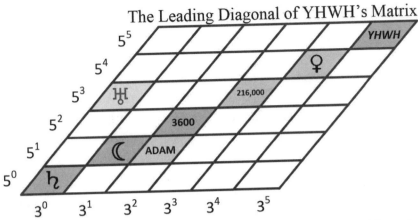

Figure 9.6. The leading diagonal of YHWH's matrix. Sixty is the balanced "way of Anu," made of 3 × 4 × 5, whose planetary gods are great: Saturn, moon, and Venus.

Further help is given from two sources: the Bible and the Olmec/Maya. Both traditions come to the limit of 144,000, Adam's name times 100 and hence 25/16 (two rows) above Adam and 5/4 (one row) short of the tzolkin. The New Testament eventually gave this Old World key, though confused by our own more exact expectation of what the cubing of New Jerusalem meant.

The number 144,000 appears in the Revelation of Saint John as

1. When 12,000 from each of the twelve tribes of Israel are sealed to form 144,000 as a final number before the Apocalypse can begin (Rev. 7:3–8), and
2. A choir of 144,000 redeemed of the earth (Rev. 14:1–5) sing a new song before the Throne (D = 144,000), to the four beasts C–G–A–E and the twenty-four elders, the remaining twenty-four reciprocal tones in this mountain.[4]

The number 12,000 occurs again in Revelation 21:16 as the number of units in one side of the city of New Jerusalem, which is a cube. Were it literally a cube with that side length it would be a vast harmonic limit of 1,728,000,000,000, 200 times bigger than the flood-hero limits of 8,640,000,000 (chapter 2), yet of the same power of three, $3^3 = 27$.

THE HEAVENLY CHOIR

The new understanding of harmonic astronomy seen in Revelation (ca. 200 CE) was that (in the octave limit 1,440) 1,000 (Uranus) is the cornerstone of heaven, transfigured from earth's cornerstone (Saturn = 1,024) by a minor diesis of 125/128. This allowed 1,000 (25/24 of the lunar year = 960) to be considered equal to 1, a unit albeit divided into 1,000 parts, each a matrix unit. This new cornerstone enables heaven to function as a single whole, in the same pattern as earth but for a different planetary context, the inner solar system, a "flying serpent." In Revelation the city was measured using a golden reed.

> And the city lieth foursquare, and the length is as large as the breadth: and [the angel] measured the city with the [golden] reed, twelve thousand furlongs. The length and the breadth and the height of it are equal. (Rev. 21:16).

New Jerusalem's side length of 12,000, equal to the number "sealed" in each of the twelve tribes, can now be seen as twelve repeated within the new heavenly creation that (divided by 1,000) repeats the World Soul of 6:8::9:12, as shown in figure 9.7. There are 150 lunar years and

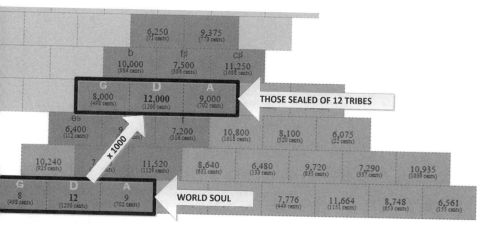

Figure 9.7. Raising the World Soul. In chapter I the simplest octave was formed using 6:8::9:12 and called the World Soul. Here the new cornerstone of 1,000, the angel's golden reed, can measure it out as 12,000 to be the sides of the city.

144 Uranus synods in 12,000 matrix units (the 24/25 relationship of the lunar year to Uranus being 144/150 = 0.96 = 24/25). The wall of New Jerusalem was also measured in a different way: "And he measured the wall thereof, an hundred and forty and four cubits, according to the measure of a man, that is, of the angel" (Rev. 21:17).

This measure is called both the measure of a man (144)* and also of an angel (Uranus = 1,000). Thus 144 Uranus synods equals 12,000 matrix units cross-referencing the new usage of the transfigured units in which 12 is the measurement and 1,000 is the measure, a golden reed "like unto a rod" (Rev. 11:1). One thousand matrix units divided by the lunar month of 80 units is 12.5 lunar months, the synodic period of Uranus, and so the golden reed is an astro-harmonic measure.

The side length and wall have set the scene for a new choir of 144,000 and a new song for the sons of Adam. Applying the same rationale, of replacing 1,000 units with 1 unit, the Revelation choir of 144,000 becomes 144 units, the ideal pentatonic limit of 72:81:96:108:128:144, as seen in figure 9.8.

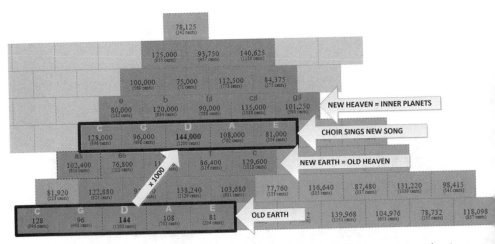

Figure 9.8. Composite for 144,000. The large number 144,000 is repeating what is on earth (144) in heaven in all but the zeros (× 1,000). These sons of Adam are 100 (two rows) above the great man as 1,440.

*The ten in Adam's 1,440 being the spirit of God.

But unlike the limit of 144, the choir have three 10s with which to extend their region of symmetrical tones above and below the pentatonic register. The choir are now singing under a new chromatic heaven (the inner planets) and a new chromatic earth (the old chromatic heaven of Adam = 1,440).

In the old heaven of 1,440, the upper row used to contain e, b, f#, and c#, but the new earth of 144,000 now contains a♭, e♭, b♭, f, and c, where the a♭ tritone has become symmetrical, while the old heaven has been chromatically flattened (by 25/24) to become the new heaven. The extra tritone of a♭ marks that the redeemed choir is able to see good and evil without doubling 1,440 to 2,880. The new heaven has g# added to it: e:b:f#:c#:g# as the "feathered serpent"—that is, Mercury synod : eclipse year : tzolkin : Mars synod : Venus synod. The diagrams of scales in figure 6.7 (page 125) around the limit 2,880, within the bed of Ishtar, now apply to the choir of 144,000, but without exceeding or perhaps then *transgressing* the identity of man in God's image and likeness. While the old heaven was sublunary, with the moon and sun dominating the lower heavens, now the inner solar system is harmoniously the new heaven for the choir.

NEW JERUSALEM

The same logic, applied once more, multiplies 144,000 by 12 and obtains New Jerusalem as 1,728,000, a million less than the true cubing of 12,000 gives; instead, a combination of the cube of 12 and the cube of 10 or $120^3 = 1,728,000$ (see figure 9.9). In Babylon this number was the ark of the Universal Flood,[5] while in India it was the Krita Yuga or Golden Age.[6] If so, then the ideas of Jerusalem and of the ark were conflated in forming New Jerusalem, a city to survive a flood by flying above the waters, a city representing God's spiritual government of human society according to harmony, with both the inner and outer planets, above and below.

Contriving to number New Jerusalem through using an irregular meaning for the cube of a number (where 1,000 = 1 unit), cubing of the side length produces $12^3 = 1,728 \times 1,000$. A similar "abuse" of cubing

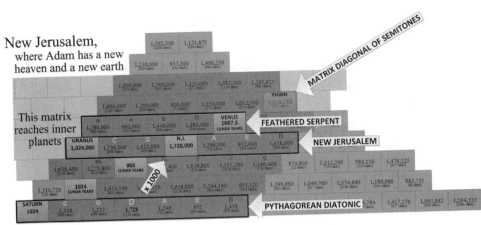

Figure 9.9. Composite of New Jerusalem.

was applied to the Babylonian ark: specifying 3,000 gallons of bitumen (3 sars), 3,000 of pitch, and 3,000 units of oil to also equal 1,728,000. The limit is then composed of primes $2^9 \times 3^3 \times 5^3 = 1{,}728{,}000$: since $3{,}000 = 2^3 \times 3^1 \times 5^3$ one can call $5^3 = 125$ unity and form $3{,}000/125 = 24$. This indicates that 1,728,000 can be $24^3 \times 5^3 = 1{,}728{,}000$, the size of the Babylonian ark. Ernest McClain diagnoses modernity's problem in dealing with such archaic forms of numeracy:

> Western theology has not even begun to ponder the riddles it inherited from the slow accumulation of adequate mathematical symbols. Again and again ancient cosmologists resorted to very thinly disguised rational meanings which have become totally meaningless as we have progressively forgotten earlier modes of thinking.[7]

And it appears New Jerusalem was a return to the Babylonian ark, the precedent for Noah's ark that in remodeling extant Babylonian flood stories, recalled the heroes who had rescued mankind from a flood. The Sumerians (prior to 2500 BCE) appear to have first developed this story, though it could have come from the Indus Valley. One can see the heptatonic Babylonian ark rising, like a ship, floating on a flood caused by three 10s, so as to sit on the leading diagonal, a spine of powers of 60, being 8×60^3. The Sumerian ark is located in the same

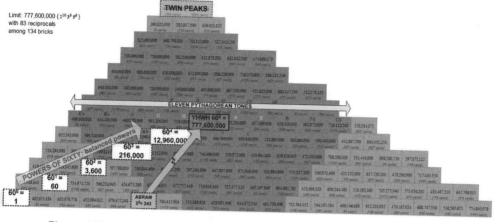

Figure 9.10. Some features of YHWH matrix. There are eleven Pythagorean tones, just short of evoking the discord of the Pythagorean comma Marduk wanted to kill. YHWH belongs to the progression of Anu's middle path of powers of 60, exceeding Ishtar/Venus, the fourth power, with his fifth power. YHWH stands over Abram's letter number of 3 to the fifth power and he has the classic ruling sign of mountain gods, twin peaks.

location but is eight times smaller. All of the Indian yuga lengths are also moored here, evidently being a highly desirable resting place on the waters. There are six decks, above and below; the ark was divided into (heptatonic) seven, with nine compartments (seven plus two) in this model, and seven levels of symmetrical tones. Such an ark cannot rise any higher because this is the intended terminus for human development, and one could argue for it as a religious dream congruous with a Far Eastern notion of emancipation from the mundane sublunary world in which mankind arose.

Having stacked heaven on top of earth as two rhomboid bricks, different by 125/128 harmonically but temporally by 1,000 matrix units, YHWH is the final part of the canvas: in the picture himself, yet a sidebar and essentially free of the action. Figure 9.10 shows the harmonic mountain for YHWH as 777,600,000 and the inner and outer solar system planets seen from Earth are all in the bottom left quadrant. Perhaps this is the great Purusha (soul) of India who gave only one-quarter of himself to the Creation.

1. A THOUSAND heads hath Puruṣa, a thousand eyes, a thousand feet. On every side pervading earth he fills a space ten fingers wide.
2. This Puruṣa is all that yet hath been and all that is to be;
-   The Lord of Immortality which waxes greater still by food.
3. So mighty is his greatness; yea, greater than this is Puruṣa.

All creatures are one-fourth of him, three-fourths eternal life in heaven.[8]

That Great Soul was "ten fingers wide," and hence Brahma's flood limit of 8,640,000,000, as super-god is pointed to, while YHWH is only nine fingers wide yet his "quarter" embraces both the 365-day year and the Venus synod. Anchored by ABRAM as $3^5 = 243$ but then multiplied by 5 (five times) YHWH becomes the ideal balanced limit, but what of the other corner location, opposite ABRAM, of $5^5 = 3,125$ as a harmonic root?

PRECESSION OF THE EQUINOXES

In the planetary matrix, Abram is the rotation of the Earth on its axis 360 times, an axis that is tipped in the Earth orbit to bring the Northern and Southern Hemispheres alternately closer to the sun in summer and farther from it in winter. Because of its tilt relative to the sun and planets, the Earth's axis shifts its location relative to the sun, causing the four pillars of the year (the two solstices and two equinoxes), sitting within zodiacal constellations at right angles to each other, to slowly move. The polar skies also change, the pole "orbiting" the (ecliptic) pole of the solar system in about 25,900 years. This period, called the precession of the equinoxes, is not exactly fixed in duration and is only to be inferred from observation. The ancient standard length for it was a harmonic $2^6 \times 3^4 \times 5 = 25,920$ years, while modern estimates are in the range of 25,820 +/-100 years. In cultures having twelve equal zodiacal divisions, twelve ages of 2,160 years are visualized and named after the spring (or autumn) equinoctial sign where the sun sits at equinox.*

Precessional forces are unique in that a rotating mass turns any force acting on its axis into the same force acting at right angles. Though tilting does occur, the motion of the axis is "sideways" and, as tilting continues, the axis performs a pseudo-orbit then called its precession so as to more or less return pointing in the same direction, in Earth's case 25,920 years later. Abram's rotation therefore causes the planetary forces to manifest a period of Great Time within which the precession of the equinoxes returns to the same zodiacal constellation. Based on how harmony "unexpectedly" defines planetary periods, one might expect to find precession within YHWH's harmonic matrix, thus again repeating the mythic explorations of the past that seem to have much to say about the Great Year.[9]

For us, precession is already "bound" by the Sumerian norm of 360 degrees within the circle, so that one observes precession by observations indicating that the sun, moving rapidly at equinox, has progressed at equinox by one degree in 72 years among the equatorial stars.

Three hundred and sixty degrees somewhat hides the primordial nature of the year for ancient peoples, calibrated by the 365 days in a solar year. Therefore, "day calibration" in which objects move in DAYS of angle on the ecliptic was how the planetary matrix was discovered and, within it, the matrix units based on the lunar month and year. The days in a solar year are 365.2422, in reality a number of Earth rotations relative to the sun in one orbit of the Earth. In contrast, Abram's year of 360 rotations is based on rotation relative to the *stars* and it is $12 \times 81 = 972$ matrix units (see 1,440 matrix), while the lunar year is $12 \times 80 = 960$ matrix units.

The precessional period of 25,920 years divided by the 354.367-day lunar year gives 26,715 lunar years to precession. Another approach is to see how many matrix units are in precession (25,646,836.92) and to notice that the leading three ("head") numbers are $2^8 = 256$. Through

---

*Knowing where the sun sits is also, at all times, only by inference since no stars can be seen once it has risen.

recent experience, this number is in the upper reaches where Uranus now has a cornerstone role so that, dividing by 1,000 gives 25,646 matrix units. It is the leading brick in any matrix row that, when doubled, will manifest powers of 2, so that in Uranus's realm we can expect precession to appear as 25,600, which means 25,600,000 matrix units. This translates to 25872.66 years for precession, and the number of lunar years in that period is then 26,666.673 years (one-third of 80,000 lunar years).

This location is the power of 5 to the fifth, on a par with YHWH in powers of 5, just as Abram is on par in powers of 3 to the fifth. YHWH as $60^5 = 777,600,000$ is divisible by the lunar year (of 960 matrix units) as containing 810,000 lunar years, while it is also divisible by the "Abramic year" of 360 rotations = 973 matrix units as 800,000 Abramic years, showing that YHWH is more than a convenient harmonic scope but is rather the directing principle for the moon-Earth system within the solar system, its master conductor, so to speak, as the common denominator of all that takes place within a harmonic system.

One can also see that during the Sumerian, Akkadian, and Old Babylonian cultures (3000–1800 BCE) the properties of 60 as a base enabled the Venus number of $60^4 = 12,960,000$ to function as a highest

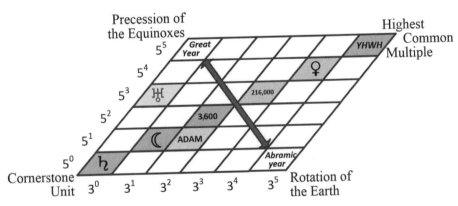

Figure 9.11. Rotation and precession of Earth within the YHWH matrix. The location of Abram is at 360 rotations of the Earth and that is related harmonically to the precession of the equinoxes, in the opposite corner of the parallelogram, as well as actually due to the gravitational perturbations on the rotating Earth by the solar system causing the Earth's poles to shift in a large circle around the polar stars.

number for the purposes of harmonic calculation. Richard Dumbrill's *Four Texts from the Temple Library of Nippur* demonstrates the existence of a complete table of "reciprocals" of 60^4 in which 60^4 is treated as 1 and the number field employed as a table of harmonic divisions, numerically unifying the two multiplicative worlds of multiple and submultiple to reduce the labor of future harmonic calculations.

This has astro-harmonic significance since the ratio 80/81, found between the lunar and Abramic months, cannot be realized by any number failing to contain $3^4 = 81$ and $16 \times 5 = 80$. This means $80 \times 81 = 6,480$ is the highest common multiple of 80 and 81, and $60^4 = 12,960,000 = 6,480 \times 2,000$. The additional 2,000, besides achieving a power of 60, was probably invoked so as to investigate just intonation above the head of Adam, above the second row of powers of 5. It is clear that just intonation required a numerical multiple of 1,000 that could raise the first three rows of "earth" up to "decimal heaven" based on a Uranus synod of 1,000 matrix units.

Ishtar seems to have become the notional leader of the harmonic work for ancient Near Eastern harmonists, but she could not provide her full bed in heaven, required for modal scales, to the Feathered Serpent. The serpent's required tritone (relative to the tzolkin period) of g# turns out to be the 365-day year of the Egyptian and Olmec civilizations. Being the nearest whole number of days within the solar year (Earth orbital period), it was one-quarter of a day shorter and hence cannot stay in contact with the seasons but drifts over 1,500 years between when it coincides with the seasons. But the 365 has great astronomical utility. When counting days, the 365-day year travels forward and backward with no fractions while also being a very precise 5/8 of the Venus synod, allowing Venus to be easily tracked on the 365-day calendar. (This technique of using whole-day years was the basis of the Roman calendar, named after Emperor Julian, which has a 365-day year.) But 5 × Venus will exceed Venus/Isis *harmonically* (perhaps as children should). Interestingly, Horus-Ra is evidently the sun as 365 days, as is clearly shown in the iconography of figure 7.8 (page 164), where the sun is above Isis, who gives birth out of her ankh.

Technically therefore, YHWH is a necessary extension of 60^4 by the additional power of 60 needed to resolve the whole of the flying/ feathered serpent. If the ancient Near Eastern harmonists had established 60^4 as their tool for navigating the sea of harmonic matrices, it appears the descendants of Moses, trained in Egypt, revealed or innovated YHWH as an absolute necessity for the whole of the flying serpent to be captured—therefore assuming that they knew of YHWH because they knew of the flying serpent. Since the late Babylonian writers of Genesis and Exodus chose to have Abram emerge from Ur of the Chaldees (i.e., what was Sumeria) and had Moses emerge from Egypt (chapter 3), then their message appears to have had composite roots, their language also being derived from Canaanite and their angelology resembling that of Persia. Had the Olmec received their tradition from Egypt's New Kingdom, and could Moses have also received it from there? And if arks were parked in heaven, below the flying serpent, by Old Babylonians, did they know the gods were up there?

10

The Abrahamic Incarnation

The Bible is probably the greatest *document of completion* for the harmonic doctrine, a doctrine that says "as in [harmonic] heaven, so on earth." Within the line of Jewish prophesy, there was a destination toward a messiah "in the line of Abraham." Before Abraham was Adam, a beginning given to a history ending with twelve tribes and, according to early Christians, the incarnation of Christ. Alternatively, a Jewish messiah would arise to fulfill the development of Adam through Abraham, incarnating as a super-prophet. This Abrahamic* idea of a completion, in the form of incarnation, can be interpreted harmonically alongside the other identities already found on harmonic mountains.

The gods of the ancient Near East were largely absent from the biblical narrative. Baal-Marduk had predated and exceeded YHWH in powers of 5, to awaken just intonation from the slumbering giant of the Pythagorean heptatonic scale. Planetary harmonics were sublimely encrypted as harmonic numbers to avoid the notion of astrology and divination, condemned within Bible narrative. Egyptian influences were indicated obscurely through the privileged education given to Moses by Egyptian royalty.

---

*Islam has no such notion because the Prophet Muhammad was the last prophet, and any direct transmission outside of his existing words is rejected by Sunnis.

The Incarnation and Trinity Implicit within Abram's Elevation

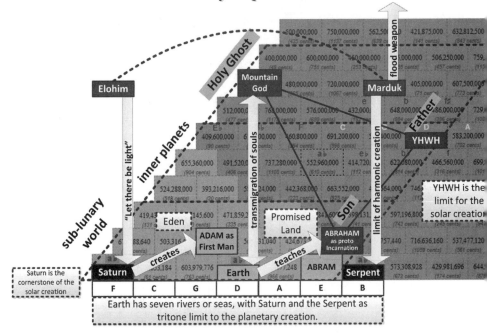

Figure 10.1. The Holy Mountain of YHWH in the context of biblical and Near Eastern harmonism. There are three tritones: the virtual Elohim to Saturn's cornerstone, God on the Mountain (Osiris) to the heptatonic Earth, Marduk to the serpent of Eden. Abram raised up by 5 becomes the savior, descended from both YHWH (limit for planetary creation) and God on the Mountain (psychopomp of the dead). The savior has the three harmonic rows of D for YHWH but "on earth."

Bible Timeline Leading to an Incarnation of God on the Mountain

Figure 10.1 shows flood heroes, located at D = 8,640,000,000, limiting the serpentine development of harmony along the base of the mountain. This limit is perfectly stated (in two dimensions) by YHWH as D = 777,600,000. The harmonic creation came from outside the solar system, conceived, we are told, by the Elohim (the reciprocal of Marduk), who founded it on the cornerstone of Saturn. Adam is then created and *elevated* out of the clay of earth into the Eden of just intonation. The first lord god was El Shaddai, a "mountain god" located as Osiris, with power over the Earth coming from a perfect tritone to himself, as with Marduk and the Elohim. In chapter 7 (see "Osiris as Resurrected Fertility God"), Osiris was presented as a ghost ruler. He awaits an incarnate representative on the heptatonic Earth—as with Pharaoh but now a descendant of Abraham. The elevation of Abram to Abraham marks the movement toward the Holy Ghost of the mountain god becoming incarnate as the son of the father god YHWH, a son who has his a♭ tritone located exactly where God on the Mountain has his (shown in figure 10.1 as "Earth"). The developed Abraham must double until he has all his father's eleven tones (see figure 10.3). His "elevation" by heh = 5 enables the incarnation to overlay all of the practical limits within the Earth and its developed modal scales (see figure 10.3).

In containing 5^7, Ernest McClain's God on the Mountain matched my reading in chapter 7 for Osiris as 40,000,000. Osiris's myth demonstrates the power of resurrection over death, demonstrated in the rebirth of his son Horus as succeeding pharaohs, and this Osiris power speaks of a third tier (rows 7 to 9 in figure 10.1: Holy Ghost) that might be the world of stars beyond the planetary world. The number field is an abstract invariant practically expressed by the planets seen through the lunar year so that, in myth, Marduk and other flood heroes project a near-perfect square root of two in preparation for a solar system that, seen from Earth, is then evolved into an image and likeness of just intonation. In contrast, God on the Mountain (40,000,000) becomes co-initiator of the Abrahamic Incarnation, perhaps the Holy Ghost of the Christian Trinity. I suggest that Abram's vision of El Shaddai and Moses's exodus from Egypt, land of Osiris, led to the hidden idea of a synthesis of YHWH and Osiris, numerically positioning Abram to Abraham, who prefigures the location of the messiah.

It is possible that the traditions that speak of a higher unconditioned source for the human spirit, from the stars or beyond, were referring to God on the Mountain and a popular Middle Eastern calendar for 360 days and an Indian one of 361 days relate the ecliptic's zodiac of constellations to this god.

THE INCARNATION'S LOST SHEEP

In chapter 7 I attributed the harmonic number 40,000,000 to Osiris, based on iconography recorded by E. A. Wallis Budge in 1911. In some of his last work, Ernest McClain equated this same number to "God on the Mountain"* allowing us to consider both these divinities within the same harmonic context.

The number 40,000,000 is on the very edge of its mountain, having no prime number 3s in its makeup, just 5s and 2s. The mountain has 101 bricks including D, and these are made up of the harmonic roots less than or equal to 40,000,000, most needing to be doubled to be greater than 20,000,000 to "be present" within that god's octave. The Rg Veda may refer directly to this mountain, as if the authors had found and then counted these brick numbers through a calculating process perhaps alluded to as "Weave back, weave forth," to sacrifice the god.

> *The Sacrifice drawn out with threads on every side,*
> *Stretched by the song of one hundred singers and one.*
> *The Fathers who are here gathered, weave their songs,*
> *They sit beside the warp and chant: "Weave back,*
> *weave forth."*
>
> *Man stretches it and man shrinks it;*
> *Even the vault of heaven he has reached with it.*

---

*McClain uses the term "god on the mountain" in different contexts within *The Myth of Invariance*, first as "Our reference tone D = 1 = n^0 is being transformed in 'numerosity' by the use of new 'least common denominators' so that we see it moved to the center of the triangle where it functions as a kind of 'god on the mountain'"(52). He often referred to the D of the limit for any given octave as its deity, but in his late work McClain specifically used "God on the Mountain" (capitalized) as the limit 5 to the seventh power but without any powers of 3 (a harmonic ghost) (79). This latter *type* of god, on the edge of its mountain, was identified and given that name.

These pegs are fastened to the seat of the Sacrifice,
They made the Sama-chant their weaving path.

What were the measures, the order, the model?
What were the wooden sticks, the butter?
What were the hymn, the chant, the recitation,?
When the gods sacrificed the god (Prajāpati, Purusa).[1]

In the Vedic tradition, Purusha plays a similar role to Adam (himself a creator of worlds through being able to name things), while Prajāpati performs the sacrifice of some facet of god that cannot exist on Earth, to manifest creation. The next lowest harmonic number to 40,000,000 is 39,585,075—a "rogue male" having no powers of two and hence incapable of being halved to form an octave limit. McClain's insight was that when Jesus referred to a lost sheep in the flock of one hundred, this is one of the hundred numbers under the octave limit 40,000,000 that must be sacrificed so that the *next lowest sheep* in the flock (the shepherd) can become D. Now on the second row, this new limiting number lies above the number of Abram (243) by the number 5, and hence it is the harmonic root of Abraham when elevated. When Abraham's root of 5 × 243 equaling 1,215 has been multiplied by fifteen 2s he becomes the shepherd valued at 39,813,120, by which time the mountain has only 99 bricks, because his limit necessitated the loss of the rogue male sheep.

If this incarnation is the Purusha of the Vedas, he approximately parallels the Hebrew Adam, who becomes a Prajāpati, as Abraham's descendent. The "song of [the] one hundred singers and one" in the Rg Veda can therefore be compared with the story told by Jesus of one hundred sheep in Matthew 18, suggesting a connection between the notion of sacrifice in Egyptian, Jewish, and Vedic traditions under the harmonic limit of 40,000,000.

For the Son of man is come to save that which was lost. How think ye? if a man have an hundred sheep, and one of them be gone astray, doth he not leave the ninety and nine, and goeth into the

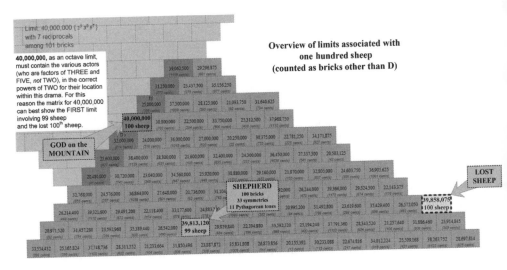

Figure 10.2. How God on the Mountain loses one of his hundred and one singers (Rg Veda, 10.130.1), aka the lost sheep in one hundred of Matthew 18:11–14, to become a shepherd.

mountains, and seeketh that which is gone astray? And if so be that he find it, verily I say unto you, he rejoiceth more of that sheep, than of the ninety and nine which went not astray. Even so it is not the will of your Father which is in heaven, that one of these little ones should perish. (Matt. 18:11–14)

Jesus's story seems a fleeting and rather obvious one concerning shepherds. The context is undeveloped but it seems a parable about Jesus being a "good shepherd" to his flock, yet implies a willingness to do more for the lost sheep than for the faithful. Considering instead the 101 harmonic singers of the limit 40,000,000 and the 100 harmonic numbers of 39,813,120 (99 without the shepherd), the missing rogue male appears as a sheep that went missing out of harmonic inevitability: the sheep went missing from the flock *because of* the incarnation of God on the Mountain or, in harmonic terms, it represents a marker for the advent of Christ. The incarnated shepherd is perhaps worried about what a human being loses in being born that he or she needs to recover. *The human is an incomplete being.* Instead of projecting onto the godhead the duty of rescuing the missing sheep, I propose this search for what is missing is the entire cause for a spiritual quest, to recover one's essential nature or "ministry." That is, perhaps one can parallel the Incarnation of God with the problems in general of human incarnation, another subject of spiritual literature after 600 BCE.

The Messiah/Christ repeats the form of the god YHWH (with a register of eleven primary tones, see figure 10.3) just below the limit of the planetary creation ($3^6 = 729$). Referring also to figure 10.1, the root of his manifestation is Abram raised up as harmonic root 1,215, a rogue male only appearing when Adam's root of 45 is reached at 1,440, alongside the appearance of the serpent or tyrant number 729. The journey between Adam and Abraham appears "mapped" onto the "range" of Adam from 45 to 1,440, at which point the journey turns to that between Abram, Abraham, and the Incarnation, mapped as 1,215 to 39,813,120, the latter limit then correlating 40,000,000 and 777,600,000 as three parts of a whole action, resembling the Christian Trinity, possible for the elevated location of Abraham. A further doubling of 39,813,120 will "leave the flock" where they are safe and "find the lost sheep" while also restoring God on the Mountain above, in heaven, as 40,000,000.

The whole matter can be confirmed within the mountain of YHWH (777,600,000) where the harmonic root of Abraham (1,215) has only just appeared as a symmetrical (wetted) brick on YHWH's mountain—any lesser limit could not embrace the incarnation in that way. The Christ has five symmetrical tones to his right, (D) A, E, B, F#, C#, which are also symmetrical for YHWH. Christ's left-hand symmetrical tones, E♭, B♭, F, C, G (D), are not symmetrical for YHWH but are "in the world" with full powers of just intonation to match those of the planetary matrix, especially the pentatonic matrix of Adam (1,440) and heptatonic of the Vedic scale (4,320:8,640, as extrapolated earlier from Curt Sachs in chapter 6).

In figure 10.3, Osiris as God on the Mountain has a♭ identical to that of Christ, outside the former's symmetrical zone but within the latter's, signifying their power on the earth, now substantiated by the Incarnation. The Trinity is then the YHWH—Father, Christ

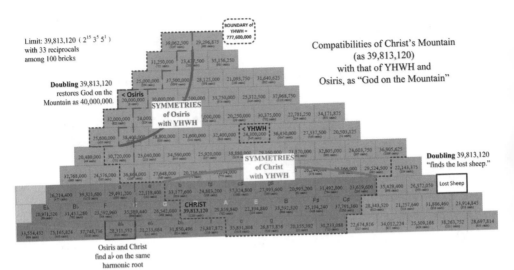

Figure 10.3. Comparative study on the mountain for 39,813,120 of the symmetrical regions of both God on the Mountain (40,000,000) and YHWH (777,600,000). The areas of symmetry within the different limits are compared in this picture of Messiah/Christ's mountain for 39,813,120. Osiris (*left*, broken line) sits within the edge of YHWH's mountain, whose symmetrical tones are bounded by the dotted line. The lost sheep Christ has to lose, to be born as a lower limit, are on the *right*. The symmetries of Christ are half those of YHWH and half of the cornerstone region on the "earth."

the Son, and Osiris the Holy Ghost—with a subject of redemption focused on the lost sheep, a sheep tonally indistinguishable from both Osiris and Christ. Incarnation is not possible without loss or sacrifice, presented in the story of Abraham as YHWH's dramatic requirement he should sacrifice his son, Isaac, who was only born due to Abraham's elevation to the location of the future Christ. One can propose therefore that the single gesture of Abraham's elevation and his willingness to sacrifice Isaac signified the future incarnation of the Holy Ghost as God's son.

ASTRONOMY INCARNATE

What do the proximate harmonic numbers 40,000,000 and 39,813,120 mean, within the world of time organized per matrix units? This can be investigated by multiplying these numbers by the matrix unit of 0.369 days to obtain a number of days. Once "on the mountain" and not within an octave, every number of days can be divided and multiplied by 2 to find if it is a multiple or submultiple of a recognizable period of time.

TABLE 10.1

Mythic I.D.	(matrix units)	(days)	scaled by	time period (days)	
God on the Mountain	40,000,000	108362210.2	2^{-11}	20 × 361 (20 × 360.5)	*Brihaspati samvatsara*
The Incarnation	39,813,120	14696311.54	2^{-21}	7 (7.0077)	Seven-day week

I propose the top three rows (figure 10.1), within which Osiris is found as God on the Mountain, belong to a domain of potential, in the universal laws, for the essence class of planet-bearing stars. This cosmic level is mediated on the Earth by the ecliptic, the sun's path through a very specific backdrop of stars during each solar year. The stars the sun passes through are the zodiacal constellations or signs. Jupiter, the most starlike planet, takes an average of 361 days to traverse a single sign, a periodicity (to this day) only noted by Indian astronomy, where it is called the *Brihaspati samvatsara* or Jupiter period.* The actual period would be 360.5, thus also perhaps explaining the Babylonian calendar for 360 days as having been more than an abstracted solar year and rather a similar approximation to God on the Mountain.

Jupiter is already dominant in the role of holding the twelve-lunar-month year in an 8:9 relationship to its synod, a lunar year that forms the basis of all the harmonic relationships to the other inner and outer planets. God on the Mountain connects the average motion of Jupiter

---

*Jupiter is the visible component in a complex set of relationships involving the solar year (the sun and Earth) and Saturn's trigon with Jupiter every twenty years.

(in 361 days) with each of the twelve zodiacal constellations in turn. The angle swept out by Jupiter in 361 days is 360 degrees divided by 12 equaling 30 degrees of the zodiac. Twelveness as a wheel (or chakra) was and is a very important symbol for patriarchal societies of the Hebrew, Vedic, and Indo-European* peoples, while twelve tones are the twelve chromatic tones of the octave, allocated within diatonic scales as five tones and two semitones. If God on the Mountain makes Incarnation possible, then this Brihaspati calendar is regulating the cosmic world (the zodiac) relative to the unique possibility on Earth for individual human soul-expression in life.

The Incarnation number (39,813,120) can similarly be reduced to the 7-day week, a week equitable harmonically with the ancient "year and a day," a framework of time associated with the male role in matriarchal societies. A male king could rule for 364 days, to be ritually killed on the 365th day, paralleling the death of the sun and initiating the appointment of a new king. The 364-day calendar is commensurate with the 7-day week, a week presumably extant when the Bible was written around 600 BCE and presenting the Elohim as creating the world in 7 days. The entirety of Indo-European peoples named these 7 days after the sun, moon and planets (as we do in English), while the Jews preferred to number their days. Seven divides into the synods of Saturn (378) and Jupiter (399), leading to 364 days being called by Robert Graves the Saturnian year, Jesus's portrait having become settled as Saturnine and Indo-European rather than resembling Apollo in early centuries.

The 364-day year was written into the souls of Europeans because mothers in a matriarchal society protect, name, and bring up sons irrespective of who fathered them. Long gone, the pattern of the matriarchal world is a suitable metaphor for the cosmic world of Incarnation, where God is your absent Father and Earth is your Mother. Per Robert Graves, the system of matriarchy became weakened by the ability of kings, through their authority, to live longer than 364 days by substituting another to be sacrificed, extending their reign to sixty "years,"

---

*The name Zeus for Jupiter is Indo-European.

perhaps meaning sixty lunar months or five years.[2] This seems the case with saturnine god Kronos in Hesiod's *Theogony:* he swallows his own children lest they replace him, only to be foxed by the earth mother Rhea, who, tiring of this performance, saves Zeus/Jupiter, who ushers in full patriarchy, at which point the connections to the past fall into myth and ritual, along with the 364-day calendar and meaning of Incarnation as relating somehow to the 7-day week as sacred.

The exact antiquity of the 7-day week is contested, but it is a great calendar when expanded to 364 days with 13 × 28-day months, resembling the 260-day calendar in Central America as thirteen periods of 20 days: these two calendars forming a tritone of 28/20 = 7/5 or 1.4 to one another. The 260-day calendar runs the numbers 13 and 20 against each other to make a repeat pattern of 260-day names (the tonal or day-name by which a person is still typified by Mayan peoples at birth). The benefits of the 364-day calendar divided by a 7-day week include the fact that all events in one year are on the same day of the same week and month (of 28 days) on every further year.

The 364-day calendar, keeping sacred days on the same day, after 49 such years resets itself to the solar year. After 49 years of 13 months, or 637 months of 28 days, one could add 2 more months giving 639 months that, multiplied by 28 days per month, and divided by 49, generate 365½ days, a slippage of just ⅒ day every 49 solar years. This makes the Jubilee calendar able to track the seasons to high accuracy by counting forty-nine 364-day years and adding 2 more Jubilee months "out of time," months during which slaves were freed and debts cancelled. The Jubilee is described Michael Segal,

In contrast to the lunar-solar calendar found in Rabbinic sources, Jubilees followed a solar calendar of 364 days per year, to which it refers as a "complete year" (שנה תמימה):

Now you command the Israelites to keep the years in this number—364 days. Then the year will be complete and it will not disturb its time from its days or from its festivals because everything will happen in harmony with their testimony. They will neither omit a day nor disturb a festival. (6:32)[3]

A similar calendar has been discovered in some of the Dead Sea scrolls from Qumran, but its oldest provenance is probably the Book of Enoch, whose date is in the range of 150 to 300 BCE, where the 360-day year of Babylonian astronomy is mentioned before the 364-day calendar.

Another utility of the complete or whole year is that it is accurately 12⅓ lunar months long, enabling the synchronizing of 3 of its years with 3 lunar years plus 1 lunar month, that is, 37 lunar months and the habit of adding an extra month to 3 lunar years in 3 solar years. The lunar and 364-day calendar therefore ran well in parallel, but also $6 \times 37 = 222$ lunar months, 1 lunar month short of the 18-year Saros eclipse cycle of 223 months. It is therefore a potent system for counting time for both astronomical and sacred purposes, while retaining contact with the lunar year. A Jubilee period of 49 years (7×7) came to be thought of as 49 solar years, as the Jubilee period became a memory.

The Jubilee also has a relationship to the matrix. The 2 (competing) months, the lunar of 29.53 days and the 364-day calendar's of 28 days, would have notably synchronized every 540 days,* while 4 solar years of 365 days are 1,460 days long. The number 540 divided by 1,460 gives the matrix constant of $27/73 = 0.369863$ days. This number correlates best with the synod of Uranus,† and when God on the Mountain of 40,000,000 is divided by 625, it relates to that synod. This gives us a glimpse of the two gods of the Trinity bringing about the Incarnation since God on the Mountain is an exalted Uranus so that the ratio of 540/1,460 is a ratio between the two calendars: God on the Mountain (364 days) and YHWH (365 days). The Incarnation is of time seen from Earth interpreted by the sensorium of living intelligence as a message of harmony as to ultimate destination of all sacred efforts.

---

*This is computed by taking the difference between the two proximate time periods, $29.53 - 28 = 1.53059$ days, and dividing that difference into the shorter to normalize two periods as an $n/n + 1$ (superparticular) ratio, in this case 18.29:19.29. If n is then multiplied by the longer period (29.53 days), the result is 540.22 days, as also when $n + 1$ is multiplied by the shorter period (28 days).

†Which is $1,000 \times 1/80$ of the lunar month, that is 29.52/80 days (0.369132 days = the synod of Uranus divided by 1,000).

The Messiah was therefore associated with Jubilees, an institutional form of redemption from debts and slavery. The 364-day year was probably used by most ancient astronomers in the same way modern astronomers use the Julian day system (of 365 days) and the Olmec/Maya the 13-day and 20-day weeks combining to a sacred calendar of 260 days in the tzolkin. One needs a system of pure counting of days to reach into the past and into the future by keeping records of events and their periodicities. Incarnation was therefore an ancient calendar held sacred and associated with the death of a male king, making for the story of crucifixion for a Jewish messiah, alongside the periodic redemption of debts and slavery.

If God on the Mountain is the Brihaspati year of 361 or, at 40,000,000 matrix units, twenty such periods, then, in contrast to the Jubilee calendar, the god Osiris became linked to the origination of souls, their transmigration and adjudication in the afterlife. Osiris was tricked by Set (Saturn, the cornerstone) into lying in a coffin *made to perfectly fit him*. The lid was slammed shut in the first of many indignities, though Isis managed to eventually find his remains (scattered like those of John Barleycorn, and like Hesiod's Uranos, whose genitals were cut off to make Aphrodite/Isis), reconstruct, and magically resurrect him as a son, Horus, who defeated Set (as should Hamlet have defeated his uncle). The story of Osiris's life was a story of incarnation as incarceration (as a body) and then dismemberment (as the loss of structures of meaning). Osiris, Isis, and Horus represent three phases of Incarnation.

To give the Earth and human life a harmonic context, the solar system was made to attain the necessary relationships of sun, moon and planets when seen from the Earth. These are not merely harmonic but define the solar system as *the smallest numerical system possible to achieve just intonation*.

The two now-unusual calendars of the 360-day/361-day year and the 364-day year are each powerful calendrical tools relating to the astronomical exploration of time rather than of space. Harmonic gods extend beyond being planets as such, to being calendrical tools made intelligible using the harmonic matrix. This should have been expected, as it is explicit in the Olmec tzolkin, but our *astronomy of space* has displaced an ancient *astronomy of time*, making the ancient gods incomprehensible. Instead of using a single calendar, each discovered celestial interval of significance creates another calendrical duration, relating to other calendars, to form a (rational) temporal pantheism by which they can be remembered. In its wake, the Christian Trinity, perhaps unconsciously, referenced three such calendars for 361 days, 364 days (by implication), and 365 days, these being God on the Mountain, the Abrahamic Incarnation, and YHWH, respectively; a threesome forged out of the elevation of Abram to Abraham, after Adam's necessary fall. Abraham never refers to God as YHWH, but as El Shaddai—the mountain god—while Moses explicitly refers to God as YHWH, and exits the land of the pharaohs, each pharaoh a son of Osiris, resurrected.

DISCOVERY OF URANUS

Uranus, seventh planet from the sun, performs an unexpectedly significant role in the harmonic matrix. Its synodic period divided by the matrix unit of 0.369 days causes it to appear in Adam's matrix for 1,440, as 1,000. While the matrix unit is very close to 1/80 of the lunar month, it is even closer to being 1/1,000 of the synod of Uranus, the time between its loops as the orbit of the Earth passes that of Uranus. Relative to the lunar year, Uranus is 25/24 longer, but the possibility of detecting Uranus to measure its synod seems too difficult for ancient astronomers since the planet is close (+5.5 to +6.0) to the limit for naked-eye visibility (+6.5).

However, it is easy to see that the two calendars used by the Egyptians, of 360 days and 365 days, have a ratio between them of 73/72,* as do other periods around this number of days, when they differ by five days. The Jubilee calendar of 364 days, when 5 days are added, generates the synod of Uranus, namely 369 days. We saw in chapter 1 that the harmonic matrix was discovered between the lunar

---

*This ratio of 73/72 is the Pythagorean comma, a problem that results from the cycling of fifths in which powers of prime 3 grow until the twelve tone classes generated cannot exactly equal the octave doubling, which has to be exact.

Discovery of Uranus from calendric differentials and the harmonic matrix

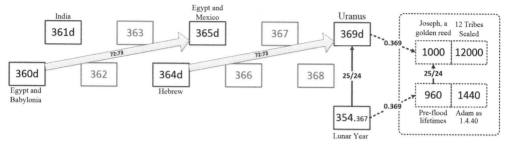

Figure 10.4. The possible discovery of Uranus using calendric differentials correlated to the harmonic matrix of limit 1,440. Two of the year lengths on the left straddle God on the Mountain within the astro-harmonic matrix, and 364 days is the calendar location for the Christ. The Uranus synod has the same Pythagorean comma ratio of 73/72 to 364 days as the Mayan haab, harmonically 2/3 of YHWH, has to Egyptian year of 360 days. The right side shows how Uranus to the lunar year chromatic semitone prefigures the angelic reed used in chapter 9 to develop the harmonic narrative of Revelation.

year, Jupiter (fifth planet), and Saturn (sixth planet) by comparing those planets to the lunar year, yielding two of the primary intervals used in practical music: the tone of 9/8 and semitone of 16/15. When this period of 369 days is compared to the lunar year, the ratio found is the chromatic semitone of 25/24, and on the matrix for 1,440 this ratio is found (where all chromatic semitones are) two rows above, as 1,000 above the lunar year as 960, which is beside Adam's D of 1,440 (see figure 10.4), equaling eighteen months.

This indicates that while observational astronomy led to discovery of the harmonic matrix, the matrix could in turn point to time periods not yet observed, giving a fine estimate for their length.

Intelligent Star Systems

The *harmony of the spheres* can only be found in our world of time, where it is a strong and compelling phenomenon. Such a harmony was no prescientific fantasy. Pythagoras, who coined the term, probably did so based on the geocentric time world, a view lost to history apart from cryptic references that can no longer be interpreted.

In our age of system science, musical harmony is not thought relevant to the design of dynamic systems such as the planets, yet they appear adapted to just intonation seen from the exclusive perspective of our planet. Why should our planet have a harmonious view of time, and what difference does time's harmoniousness make to life on Earth? Is there some other purpose to this harmony—or none at all? To answer such questions one has to recognize just intonation as being a holistic system that demands human insight into the nature of whole phenomena (a so-called gestalt). Such gestalts flow from the need to see higher-level relationships rather than the raw complexity of their parts.

All higher structures of meaning subsume lower levels of meaning. For example, microclimates are a structuring of meaning higher than trees, water, weather, and topography, usefully integrating these parts within a newly perceived whole. Such insights reveal a higher idea that indicates new potentials within a system. The new level of conceptual order has not changed in the phenomenon but how we relate to it. This profound faculty is the basis of what we call understanding rather than knowing, and it enlarges our "world." The world is already structured, and a sensory insight re-creates that structure as a simplifying aspect already present, to expand the intelligibility of the sensory world and with it, our present moment. Insight and the world's creation were considered similar acts within ancient cosmologies, in that an insight about the world resembles the structure of the world as it would be conceived by any god in the act of creating it. Such a vision involves a special effort but provides a creative view of the world, in which simplicity and relatedness replace functional complexity with a new appreciation of the sensory world. The celestial behavior in Earth's skies is a prime example of such an action: the rotation of Earth, its orbit around the sun, the moon's orbit, and its illumination by the sun complicate the observed orbital periods of the other planets and yet, that added complexity has produced harmonic simplicity between synodic periods!

Chapter 1 showed how Late Stone Age astronomers used geometrical counts of synodic periods to discover this harmony of the spheres, which modern astronomers have not seen because scientific calculation methods deal instead with planetary dynamics modeled by equations. Simplicity has somehow adapted our solar system without breaking physical laws. At the level of gravitational dynamics, many complexities were required to achieve just intonation seen only from Earth, especially the lengthening of the lunar month as an intermediary to the planetary synods seen from Earth. Any demiurgic preference for harmony (seen from Earth) resembles the human gestalt that revealed the harmony of the spheres to human sensory intelligence in the Late Stone Age, and it must be noted, humanity has become demiurgic since the Stone Age, creating man-made worlds.

Demiurgic intelligences are probably part of each star system and, if our star has a demiurgic intelligence, this action seems to have used the moon to establish a justly intoned time world for the third planet. It adapted the unchanging orbital pitches of an *n*-body planetary system to present harmonic synodic systems that planetary orbital periods alone could never express. Our geocentric system is harmonically founded between 1 (the Saturn synod) and the fifth power of 60 (YHWH, as 365-day year), which is the smallest *numerical resolution* to contain just intonation of both inner and outer planets, as in the implied holy mountains of our ancient texts.

Astronomical Periods and Their Matrix Equivalents

In the harmonic matrix, powers of 2 are not defining, while combined powers of odd (male) primes 3 and 5 (the patrix root) allow relativity to a standard matrix with D = 1,440 matrix units (MU), even when their patrix exceeds that limit of 1,440. This allows the whole range of astronomical periods to be viewed as a continuum, within which octaval entities (as in holy mountains) have their own meaning.

In the chart that follows, we

a. define tonal values (if appropriate)
b. name the period
c. give a definition
d. define its actuality, in days, years or MU
e. identify its patrix root value ($3^q \times 5^r$)
f. give the patrix in time units
g. show the powers of 2 required to achieve the period
h. give that period as a matrix ideal value
i. show the error ($h - d$) as a time period

When a column's units or meaning changes vertically, that is noted. For example, matrix super gods has (d) in MU and (h) in lunar years; periods under the Feathered Serpent offer (1) as an ideal length in 13-day weeks. This table is not exhaustive but represents what is considered in the main text.

tone	name	definition	actual (days)	harmonic root	length (days)	powers (of two)	harmonic (days)	error (days)	Ideal (÷13)	Jupiter (synods)	lunar (years)	Rotational (years)
	lunar month	days between same relation to the sun, seen from Earth	29.53059	5	–		–	–				
	matrix unit (MU)	one-eightieth of lunar month	0.369132375	1	–		–	–				
Just Intonation (Earth)												
ab	Saturn Synod	days between same relation to stars, seen from Earth	378.09	1	0.36913	10	377.992	-0.098				
G	Lunar Year	days for whole number (12) of lunar months in year	354.367	15	5.53690	6	354.367	–				
D	Supplemental Year	days for 18 lunar months	531.551	45	16.61096	5	531.551	–				
A	Jupiter Synod	days between same relation to sun, seen from Earth	398.88	135	49.83287	3	398.663	-0.217				
	Rotational Year	days for 360 rotations of the Earth	359.022	243	89.69917	2	358.797	-0.225				
Feathered Serpent (Heaven)												
ab	Uranus Synod	days between same relation to sun, seen from Earth	369.66	125	46.14155	3	369.132	-0.528	–			
C	Mercury Synod	days between same relation to sun, seen from Earth	115.88	625	230.70773	-1	115.354	-0.526	117 (9)			
G	Eclipse Year	days between eclipses of the moon, at same node	346.62	1,875	692.12320	-1	346.062	-0.558	–			
D	Tzolkin	Twenty weeks of 13 days	260.00	5,625	2,076.36960	-3	259.546	-0.454	260 (13)			
A	Mars Synod	days between same relation to sun, seen from Earth	778.94	16,875	6,229.10882	-3	778.639	-0.301	780 (60)			
E	Venus Synod	days between same relation to sun, seen from Earth	583.92	50,625	18,687.32648	-5	583.979	0.059	585 (45)			
g#	Haab	number of whole days in solar year	365.00	253,125	93,436.63242	-8	364.987	-0.013	364 (28)			
Super Periods												
	Precession	est. years for equinox in same relation to the stars	25,872.6666 (years)	3,125	9,449,788.8	13	25,872.6642 (years)[2]	0.868850[5]				
Matrix Super Gods			_matrix (MU)_							Jupiter (synods)	lunar (years)	Rotational (years)
Marduk, Indra, etc.	Universal Flood		8,640,000,000	2,109,375	778,638.6035	12				8,000,000	9,000,000	8,888,888.9
Osiris	God on the Mountain		40,000,000	78,125	28,838.4668	9						
YHWH	God of the Bible		777,600,000	759,375	280,309.8973	10				720,000	810,000	800,000

Ancient Use of Tone Circles

Reunification of Tuning by Number with Tuning by Ear through Reason and Visual Symmetry

This discussion seeks to explain how, to quote Ernest G. McClain, "scribal curiosity, a modicum of experience with a 'straight-edge and rusty compass,' and perhaps also the development of base-60 arithmetic in the fourth millennium BC" could have deduced the twelvefold nature of the harmonic world and placed tones correctly in their near-semitone distributions, per "hour," on a "clock face." This would then correspond to the subject's logarithmic perception, by ear, which is unlike the manufacture of tones by an instrument's dimensions. Tones and intervals within an octave will grow or diminish in frequency logarithmically, while physical string lengths will grow or diminish according to their linear lengths, these being anciently modeled arithmetically by integer values and the rational fractions between them. Only by returning tonal string lengths and interval ratios to the logarithmic world of tuning by ear could the symmetry and twelvefold nature of tonal matrices be unified by eye and mind and revealed as the species presented in Ernest G. McClain's *The Myth of Invariance*. We here describe a simple procedure that could have made this possible in the fourth millennium BCE, through the contemplation of the symmetrical rising and descending intervals found within an octave. This approach reflects on other features found in the later tuning theory of Plato, such as the preference for symmetrical tone sets, the attention given to the octave's geometric mean, and the recognition of symmetry within the ear's logarithmic world, and reflects the duality of the octave's bounding Ds, as "ambidextrous twins," to either side of which is then a tone circle.

The octave defined by a range of string lengths between 360 and 720 is perhaps the most ubiquitous example given explicitly by Plato or found implicitly in textual allusions in classical and ancient texts. It appears to be modeled on the solar year of 365 whole days or 365¼ days, and 360 was used in the calendars of many cultures, most notably in Egypt. It is the factors of 360 as a number that made it more suited to subdivision than the solar year and hence to its application within harmonic calculations where numerical tuning theory generates tones using rational ratios containing only factors of the first three prime numbers, 2, 3, and 5. The same number of 360 subdivisions was also adopted by the Sumerians for division of the whole circle, in practice the 360 degrees of the horizon, again because of its far greater utility in calculation, especially because of their use of a sexagesimal base-60 notation of numbers, meaning that 6 counters of 60 make 360, and 12 make the octave 720. Links to astronomy were forged leading to the division of the ecliptic into a set of twelve zodiac constellations, inherited by the Archaic Greeks from Babylonia and its evolved (annual) ritual and farming calendar.[1]

One can therefore see that the octave 360:720 is unique in being numerically fairly small and still in the practical range for string-length experimentation using units of length less than an inch, yet its "numerosity" had useful characteristics in providing 2s, 3s, and 5s to, in theory, enable just intonation and the formation for an approximation of twelve tones. The octave itself is best seen as a geometric doubling of a square with sides 360 units long, whose diagonal is then $\sqrt{2} \times 360$ units, or about 503 units long, though irrational. A square made using

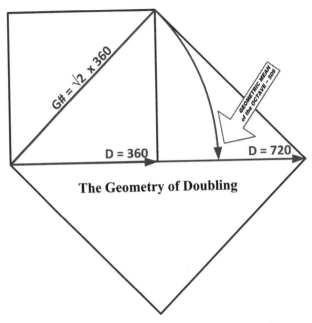

Figure A2.1. Square geometry of doubling for D = 360:720.

that diagonal as the side length then has a diagonal of √2 squared, giving us 2 × 360 = 720, as shown in figure A2.1. The √2 is the tritone of A♭ = G# and when is itself squared, the square root that contains six equal-tempered (ET) tones "adds up" to twelve ET tones because each such semitone is the twelfth root of 2, which in the logarithmic terms heard by the ear is 1/12, which added twelve times to 1 (D = 360) gives 2 (D = 720).

Methods of calculation used for studying harmonic tuning in the ancient world could not (directly) access or manipulate numbers as logarithms except through the ears as corresponding to the tones generated by strings of known length. String length numbers were integers and simple arithmetical operations—such as adding by doubling (× 2), tripling (× 3), and doubling twice (× 4) and then adding another (× 5)— and employed counters that aggregated products multiplied or divided by 3/2 (tripled then halved) or 5/4 (quintupled then halved twice) until the numbers produced "ran out" of one of the prime numbers (2, 3, 5) and so become fractional. This exhaustion of 2s, 3s, and 5s naturally

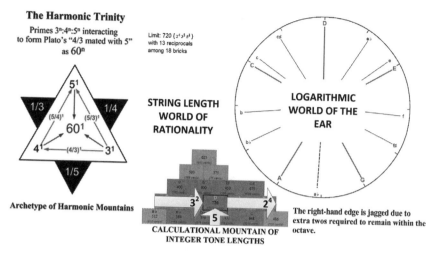

Figure A2.2. Archetype of McClain's holy mountains where limiting magnitude D = 720 has one 5, locating D one brick up; two 3s, locating D two from left edge; and two 2s from the right edge because 3:5 requires two octave resets plus three 2s, that is, five 2s, which 720 does not contain. Right graphic from www.harmonicexplorer.org online calculator.

generates an approximately triangular form, termed a *harmonic matrix, a mountain,* or, in a religious context, a "holy mountain" by Ernest McClain. At the heart of such a triangular mountain of rationality one can recognize Plato's "4:3 mated with 5," as seen in figure A2.2.

Both Ernest G. McClain and Antonio de Nicolás thought that the Sumerians and Babylonians had a way of visualizing the tone circle (shown on the *right* in figure A2.2), here produced using a knowledge of logarithms (base 2) to form angles up to 360 degrees within a "pie chart" with D at the top and the geometric mean at the bottom. McClain suggested the ancient method only needed to be *indicative* in order to have enabled important insights, hence his own words: "scribal curiosity, a modicum of experience with a 'straight-edge and rusty compass,' and perhaps also the development of base-60 arithmetic in the fourth millennium BC."*

---

*Text adapted from the draft of *ICONEA 2008,* see the following section, "Addendum: Short History of the 'Rusty Compass.'"

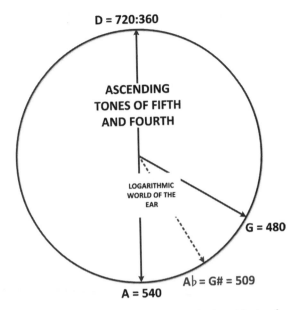

D = 720:360

ASCENDING
TONES OF FIFTH
AND FOURTH

LOGARITHMIC
WORLD OF THE
EAR

G = 480

A♭ = G# = 509

A = 540

Figure A2.3. Showing the distortions from the logarithmic of ascending fifth and fourth, and geometric mean, when string length is taken as angular measure without employing base-2 logarithms.

Is it possible that the symmetrical tone circle viewable today could have been usefully anticipated in the ancient Near East, without any rewriting of the history of mathematics for that period? I believe that there was a means for correcting the "distorted" view of linear string tones that results from translating them into angular degrees, as seen in figure A2.3, if the ancient harmonists used common sense and the octave limit 360:720.

This distortion between the linear world of number and the logarithmic world available to the ear, shown for ascending tones in figure A2.4, can be seen to be symmetrical, when displayed side by side in figure A2.4, for the same tones *descending* from D = 720.

Is it possible that an ancient experiment with a "rusty compass," and an understanding of the symmetrical nature (in figure A2.4) of ascending and descending tones, was combined with the numerical situation where the octave spans the number of degrees (360) already allocated to a circle? This would have allowed the problem to be seen "from both

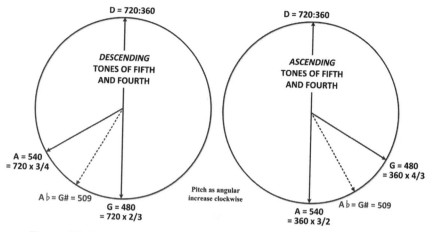

Figure A2.4. The descending (*left*) and ascending (*right*) of fourths and fifths, seen as angular, are symmetrically wrong with respect to the logarithmic world where the dotted geometrical mean should be central.

ends," that is, from both high and low D, and symmetrically. The correction required is intuitively obvious: the geometric mean should lie exactly between the descending and ascending fourths, at the bottom of the tone circle, while (as drawn) A and G should (most simply and

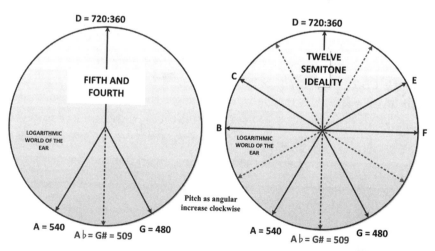

Figure A2.5. The correction (*left*) of ascending and descending movements to "be the same," placing the geometric mean below and opposite D; and (*right*) the creative leap of moving to the twelvefold semitone model as geometric consequence of the twelve-semitone model within 360 degrees.

aurally) lie equally spaced, to either side of it. If the existing distance of about 60 units between A and G is halved and used as an angular unit, A and G are then located in their equally tempered position that is only a few (modern) cents from their actual locations.

It is then a simple step to visualize the tonal world as logarithmically divided into twelve parts such that between A and G lie two such parts, equaling a whole tone of 9/8, comprehended clearly by the two string lengths of 540 and 480, each a semitone of 30 angular units from Ab/G#. Cycling further fifths back "into" the octave from D = 360:720 gives the 9/8 whole tones E and C, symmetrically around D, which can obviously be taken as two double semitones of 60 on either side of D. Moving toward D from A and G produces B and F, respectively, so as to leave two Pythagorean leimmas of 256:243 to C and E, semitones that can be then be valued at 30 angular degrees to picture heptatonic reality. The familiar octave view, of five tones of 60 degrees each and two semitones of 30 degrees, conforms to the ear's logarithmic sensibility, while, owing to the 360 degrees within a circle and the 360 units of length between high and low D for the limiting number 720, the twelvefold nature of the musical world is revealed as found within the Pythagorean "cycle of fifths," prefiguring the (numerically irrational) ideal of equal temperament.

This miracle of false accounting allows a merely functional method to present an intuitively correct realization: *that harmony has a twelvefold structure within the octave.* The *geometric* twelfth root of the 360 units within the octave is a division of the circle into twelve sections of 30 units each, each in turn 30 degrees of the circle. However, equal temperament defines twelve equal semitone roots of octave doubling that are only equal in the logarithmic world of the ear. So why is such a division about the tritone able to equate equal division of a circle with equal temperament as heard by the ear?

The two adjacent equal-tempered semitones closest to the tritone are of the average size of all the twelve equal-tempered semitones because they bracket the tritone's geometric mean for the whole octave.

This average size allows them to divide an angular circle into twelve parts while matching the G–A whole tone. Additionally, each contains

30 parts in the now conventional 360 parts of the whole circle. This means that when A and G in the octave 360:720 are seen, for reasons of symmetry, to lie symmetrically at the bottom of a tone circle, their size represents the average size of all the differently sized (equal-tempered) semitones within that octave, allowing their sectional widths to be extrapolated around the whole circle, using a rusty compass, to form a tone circle (like the one in figure A2.5).

However, neither the utility of 360 degrees nor the octave of 360:720 turn out to be essential to this, for 360 is six units of 60 while 720 is twelve units of 60. This means the G–A–D tuning order used here can be generated by a limit as low as D = 12 to display the same proportionality and, in theory, allow the above symmetrical insight to be made. Using 360 degrees usefully ties the tone circle to the associated holy mountain for limit 720.

It is simply a fact that the two equal-tempered semitones surrounding the tritone are of average size, and the 6:8::9:12 geometrical division of a circle into twelve sections will give each, on the circumference, 1 unit and that unit will then be proportional to the correct rendition of an equal-tempered whole tone on the circumference, which then may be repeated around the circle, using a "rusty compass," as in figure A2.6.

One should ask, why twelve? The reason has to do with the closeness of these two equal-tempered semitones (which together equal 1.1225 or 200 cents) to the larger whole tone of 9:8 (= 1.125 or 204 cents) just 4 cents different. It is a requirement of the human experience that only the most numerically simple tonal ratios, based only on the earliest primes (2, 3, 5), will be harmonious. In nature, this leads to an octave of five whole tones and two semitones, twelve semitones in all, and within the symmetrical view of D, a whole-tone interval must "bridge" the tritone between G and A, and this causes the closeness in value of the two (symmetrical and equal-tempered) semitones to the larger whole tone, to correlate with the symmetrical diktat of the octave. While even roots of 2 other than the twelfth are interesting, it is only the twelfth root that can provide this confluence of the logarithmic world of aural harmony and the linear world of rational string lengths, to form a circle populated, to the eye, by tonal divisions true to the ear and created through

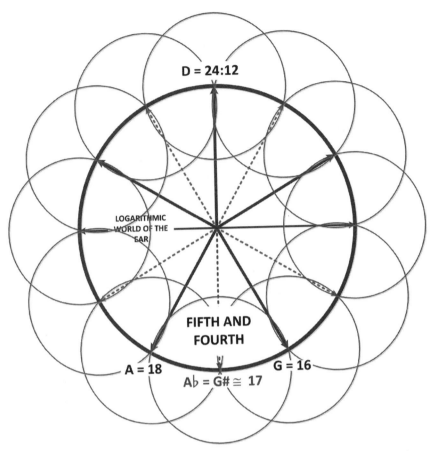

Figure A2.6. The smallest limiting octave needed to obtain a rational G and A whose distances enable the construction of the twelfth root of 2 giving twelve segments of a circle, using a "rusty compass." A limit of 24 allows the geometric mean to be very nearly 17.

good symmetrical reasoning rather than a knowledge of logarithms.

This perhaps caused McClain to ponder long the title of his proposed *Festschrift for Antonio de Nicolás,* "What is Twelveness," yet perhaps not see the exact reason for this confluence of two worlds (found opposite D within the octave throughout history), made within the number field long before the first rusty compass, but certainly a blessing for those wishing to visualize the results of their harmonic calculations within holy mountains, without waiting for our functional mathematics of logarithms. Such a shortcut is unseen because unnecessary in a modern context, but it indicates a feature of the number field similar to the golden mean of 1.618034, which reciprocates to 0.618034 and squares to 2.618034 with identical fractional excess—a feat only possible once within the number field and one leading to another human sense that sees geometric harmony in nature and ancient architecture.

ADDENDUM: SHORT HISTORY OF THE "RUSTY COMPASS"

Ernest McClain's concept of the "rusty compass" took shape over several years. In this section we present two of these incarnations.

Draft for ICONEA Conference 2008

This article, here quoted from an early draft of McClain's paper for Shuruppak 22 (Mesopotamia), "Mesopotamian Musical Imagination: Recovering Ancient Harmonic Theory from an Early Mesopotamian 'Grain Pile,'" for the 2008 International Council of Near Eastern Archaeomusicology (ICONEA) conference, is discussing the base of the mountain for the Sumerian grain pile, whose number is 1,152,000—a harmonic number as limit D = 1,152,000 that has thirteen bricks along the base of its mountain. This means the base forms a spiral of fifths up to the creation of the Pythagorean comma, so forming part of this number's significance to early tuning theory. In the published version, McClain launches instead into Inanna's bed as the compression of these thirteen tones into a throne, "a portable pattern freely moved up and down and right and left within the limits of the grain-pile or any larger matrix as a privileged subset, but eventually sealed in place by the defining limit." This led to retirement for the rusty compass, until ICONEA 2010.

> The result of this tuning procedure when "mapped" into a circle is . . . the tones realign in scale order at very nearly equal distances. The locus of the tones results from a musical fifth of 2:3 embracing seven semitone "hours" in the completed set, meaning that complementary musical fourths of 4:3 embrace five, symmetrically

opposite in the cycle. Locating either one reveals the locus of its arithmetical and tonal reciprocal. This discovery seems likely to be owed to scribal curiosity, a modicum of experience with a "straight-edge and rusty compass," and perhaps also to the development of a base 60 arithmetic in the fourth millennium BC, for the system is likely a scribal discovery rather than a musical one. The radius of any circle divides its circumference into six equal parts that conveniently symbolize six whole tones in the octave.

Publication as "Egyptian Connections: Narmer Inscriptions as Sumerian Musicology" ICONEA 2010

Here is the eventual use of the rusty compass idea, after the Pythagorean comma again came into discussion under "Computing the entire spiral of fifths system":

> A plumb line from the throne above maps the tritone square root of 2 in equal temperament at 6 o'clock, cutting the model (i.e., the first) whole tone of 9:8 in half as two semitones—never computed, and always appearing as remainders from computing consonances. All equal divisions of the octave cycle are naturally irrational roots of the ratio 2:1, but no intervals smaller than 9:8 were actually drawn, and so early musical geometry is merely illustrative, not epistemological; knowing meant knowing the right number. Within whole tones (easily mapped in equal temperament (ET) with a rusty compass in which the radius as half the diameter cuts off semitone hours along the circumference) we can approximate by eye the positions of semitone half-hours and quarter-tone quarter-hours (minimal time for observing stellar movement by naked-eye astronomy), potentially halved further by metaphorical commas not worth the trouble of defining and often subliminal to ears. Third-tones (unpopular in the West) can be approximated by two infixes within the whole tone.[2]

Notes

INTRODUCTION. THE SIGNIFICANCE OF PLANETARY HARMONY

1. Heath, *Sacred Number and the Roots of Civilization.*
2. McClain, *Myth of Invariance*; and McClain, *Pythagorean Plato.*

I. CLIMBING THE HARMONIC MOUNTAIN

1. Heath, *Sacred Number and the Lords of Time*, 44–53.
2. Heath, *A Key to Stonehenge*, 37; *Sun, Moon and Stonehenge*, chapter 5.
3. Remarked on early by Heath, *Sun, Moon and Stonehenge*, 84–85; and Neal, *All Done with Mirrors*, 69–74.
4. McClain, *Myth of Invariance*, 126.

2. HEROIC GODS OF THE TRITONE

1. McClain, *Myth of Invariance*, 76–83.
2. Based on Kramer, *Sumerian Mythology*, 74.
3. Jastrow, "Sun and Saturn," 163–78.

3. YHWH REJECTS THE GODS

1. Douglas, *Thinking in Circles*, xii–xiii, on John Myres "pedemental" writing

and Jacob Milgrom's design of a biblical ring embracing the Pentateuch and the Book of Joshua.

2. Dalley, *Myths from Mesopotamia,* 233, tablet 1 of the Epic of Creation.

3. McClain, *Myth of Invariance,* 91.

4. PLATO'S DILEMMA

1. See Rogers, "Origins of the Ancient Constellations: II, The Mediterranean Traditions."

2. My primary source for this is Bremer, *Plato's Ion.*

3. For my own starting point, see Kapraff and McClain, "The System of Proportions of the Parthenon," 203–20.

4. McClain, *Pythagorean Plato,* 2–3.

5. Levarie and Levy, "Temperament," in *Musical Morphology.*

6. McClain, *Pythagorean Plato,* 7, 178 n. 14.

7. McClain, *Pythagorean Plato,* 14.

8. McClain, *Pythagorean Plato,* 74.

9. McClain, *Pythagorean Plato,* 20.

10. Henderson, *Ancient Greek Music,* 341.

5. THE QUEST FOR APOLLO'S LYRE

1. Kilmer, Crocker, and Brown, *Sounds from Silence.*

2. Dumbrill, "Four Texts from the Temple Library of Nippur."

3. Vedic India: *Hiranyagarbha* (Rg Veda 10.121, Nicolás, *Meditations through the Rg Veda,* 226–27); Greece: the Orphic Egg (West, *Hesiod,* 205); Egypt: *Tatanen* or Primordial Hill (Naydler, *Temple of the Cosmos,* 54); China: Wu Cheng'en, *Journey to the West* and Pearce, *The Crack in the Cosmic Egg.*

6. LIFE ON THE MOUNTAIN

1. See Dumbrill, "Four Texts from the Temple Library of Nippur."

2. From https://en.wikipedia.org/wiki/Shvetashvatara_Upanishad#Chronology (accessed November 28, 2017).

3. Sachs, *The Rise of Music in the Ancient World,* 165.

4. Nicolás, *Meditations through the Rg Veda,* 157.

5. Sachs, *The Rise of Music in the Ancient World,* 165.
6. McClain, *Myth of Invariance,* 79–81.

7. GILGAMESH KILLS THE STONE MEN

1. Dalley, *Myths from Mesopotamia,* restored by Mitchell, *Gilgamesh.*
2. Mitchell, *Gilgamesh,* 91–92.
3. Mitchell, *Gilgamesh,* 172–72.
4. Wolkstein and Kramer, *Inanna: Queen of Heaven and Earth,* 56.

8. QUETZALCOATL'S BRAVE NEW WORLD

1. Milbrath, *Star Gods of the Maya,* 26.
2. Sugiyama, "Teotihuacan City Layout."
3. Linden, "The Deity Head Variants of Glyph C."
4. Sugiyama, "Teotihuacan City Layout."
5. Powell, "The Shapes of Sacred Space," 54.
6. Pool, *Olmec Archaeology,* 253.
7. Soustelle, *The Olmecs,* 166.
8. Pool, *Olmec Archaeology,* 254.

9. YHWH'S MATRIX OF CREATION

1. Derived from "Quetzalcoatl," Wikipedia, https://en.wikipedia.org/wiki/Quetzalcoatl (accessed November 28, 2017).
2. West, *Hesiod,* 6–7.
3. Christensen, *Nahum,* 30–36.
4. See McClain, *Myth of Invariance,* 110–13.
5. McClain, *Myth of Invariance,* 149.
6. McClain, *Myth of Invariance,* 74.
7. McClain, *Myth of Invariance,* 117.
8. Rg Veda, 10.90.1–3, translated by Ralph Griffiths, http://sacred-texts.com/hin/rigveda/rv10090.htm (accessed November 28, 2017).
9. See Santillana and von Dechend, *Hamlet's Mill,* where that theme is expounded.

10. THE ABRAHAMIC INCARNATION

1. Rg Veda 10.130.1, translated by Antonio de Nicolás, *Meditations through the Rg Veda*, 230.
2. Graves, *The Greek Myths*, 7.1
3. Segal, "The Jewish Calendar in Jubilees."

APPENDIX 2.
ANCIENT USE OF TONE CIRCLES

1. Rogers, "Origins of the Ancient Constellations," I and II.
2. McClain, "Egyptian Connections: Narmer Inscriptions as Sumerian Musicology," 79.

Bibliography

Bremer, John. *Plato's Ion: Philosophy as Performance.* North Richland Hills, TX: Bibal Press, 2005.

Brown, Norman O. "The Birth of Athena." *Transactions and Proceedings of the American Philological Association* 83 (1952): 130–43.

Budge, E. A. Wallis. *The Egyptian Book of the Dead.* London: Kegan Paul, 1899.

———. *Osiris and the Egyptian Resurrection.* 1911. Reprint, New York: Dover, 1985.

Christensen, Duane. *Nahum: A New Translation with Introduction and Commentary.* New Haven, CT: Yale University Press, 2009.

Creese, David. *The Monochord in Ancient Greek Harmonic Science.* Cambridge: Cambridge University Press, 2010.

Dalley, Stephanie. *Myths from Mesopotamia: Creation, the Flood, Gilgamesh, and Others.* Oxford: Oxford University Press, 1989.

Douglas, Mary. *Thinking in Circles: An Essay on Ring Composition.* New Haven, CT: Yale University Press, 2010.

Dumbrill, Richard J. "Four Texts from the Temple Library of Nippur: A Source for 'Plato's Number' in Relation to the Quantification of Babylonian Tone Numbers." ICONEA: lulu.com 2013. Available at: www.academia.edu.

Dumbrill, Richard, and Bryan Carr, eds. *Music and Deep Memory: Speculations in Mathematics, Tuning, and Tradition, In Memorian to Ernest G. McClain.* London: ICONEA, 2017.

Graves, Robert. *The Greek Myths.* London, Penguin, 1960.

Heath, Richard. "Ernest McClain's Musicological Interpretation of Ancient Texts: Musicological Narrative Structures in Biblical Genesis." In *Music and Deep Memory: Speculations in Mathematics, Tuning, and Tradition, In Memoriam to Ernest G. McClain*, edited by Richard Dumbrill and Bryan Carr. London: ICONEA, 2017.

———. "The Geodetic and Musicological Significance of the Parthenon's Shorter Side Length, as Hekatompedon or 'Hundred-Footer.'" In *Music and Deep Memory: Speculations in Mathematics, Tuning, and Tradition, In Memorian to Ernest G. McClain*, edited by Richard Dumbrill and Bryan Carr. London: ICONEA, 2017.

———. Harmonic Explorer. http://harmonicexplorer.org.

———. *Sacred Number and the Lords of Time*. Rochester, VT: Inner Traditions, 2014.

———. *Sacred Number and the Roots of Civilization*. Rochester, VT: Inner Traditions, 2004.

Heath, Richard, and Robin Heath. *The Origins of Megalithic Astronomy as Found at Le Manio*. 2010. www.academia.edu/5384545 (accessed November 28, 2017).

Heath, Robin. *A Key to Stonehenge*. Cardigan, Wales: Bluestone Press, 1993–95.

———. *Sun, Moon and Stonehenge*. Cardigan, Wales: Bluestone Press, 1998.

———. *Sun, Moon & Earth*. New York: Walker, 1999.

Henderson, Isobel. *Ancient Greek Music*. Vol. 1 of *The New Oxford History of Music*, edited by Egon Wellesz. Oxford: Oxford University Press 1957.

Jastrow, Morris. "Sun and Saturn." *Revue D'Assyriologie et d'archéologie Orientale* 7, no. 4. (1910): 163–78. www.jstor.org/stable/23283795 (accessed November 28, 2017).

Kapraff, Jay. *The Lost Harmonic Law of the Bible*. Proceedings of London-Bridges 2006.

Kapraff, Jay, and Ernest G. McClain. "The System of Proportions of the Parthenon: A Work of Musically Inspired Architecture." *Music in Art* 30, no. 1/2 (2005): 5–16. www.jstor.org/stable/41818722 (accessed November 28, 2017).

Kilmer, Anne Draffkom, Richard L. Crocker, and Robert R. Brown. *Sounds from Silence: Recent Discoveries in Ancient Near Eastern Music*. Vinyl album and booklet. Berkeley, CA: Bit Enki Publications, 1976.

Kramer, Samuel. *Sumerian Mythology*. Philadelphia: University of Pennsylvania Press, 1972.

Levarie, Siegmund, and Ernst Levy. *Musical Morphology: A Discourse and a Dictionary*. Kent, OH: Kent State University Press, 1983.

Linden, John. "The Deity Head Variants of Glyph C." *Eighth Palenque Round Table* 10 (1993): 369–77.

McClain, Ernest G. "Egyptian Connections: Narmer Inscriptions as Sumerian Musicology." *ICONEA 2009–2010: Proceedings of the International Conference of Near Eastern Archaeomusicology, British Museum, December 4–6, 2008,* edited by Richard Dumbrill and Irving Finkel. London: ICONEA Publications.

———. *The Myth of Invariance: The Origin of the Gods, Mathematics and Music from the Rg Veda to Plato.* York Beach, ME: Nicolas Hays, 1976.

———. "A Priestly View of Bible Arithmetic: Deity's Regulative Aesthetic Activity within Davidic Musicology." In *Hermeneutic Philosophy of Science. Van Gogh's Eyes, and God: Essays in Honor of Patrick A. Heelan S.J.,* edited by Babbette Babich, 429–43. Dordrecht, the Netherlands: Kluwer Academic Publishers, 2002. http://fordham.bepress.com/phil_research/23 (accessed November 28, 2017).

———. *The Pythagorean Plato: Prelude to the Song Itself.* York Beach, ME: Nicolas Hays, 1978.

———. "A Sumerian Text in Quantified Archaeomusicology." In *Proceedings of the International Conference of Near Eastern Archaeomusicology, ICONEA 2008, Held at the British Museum, December 4, 5, and 6, 2008,* edited by Richard Dumbrill and Irving Finkel, 89–103. London, ICONEA Publications.

Milbrath, Susan. *Star Gods of the Maya. Astronomy in Art, Folklore and Calendars.* Austin: University of Texas Press, 1999.

Mitchell, Stephen, trans. *Gilgamesh.* New York: Simon & Schuster, 2000.

Naydler, Jeremy. *Temple of the Cosmos.* Rochester, VT: Inner Traditions, 1996.

Neal, John. *All Done with Mirrors: An Exploration of Measure, Proportion, Ratio and Number.* London: Secret Academy, 2000.

Nicolás, Antonio T. de. *Meditations through the Rg Veda: Four-Dimensional Man.* York Beach, ME: Nicolas Hays, 1976.

Pearce, Joseph Chilton. *The Crack in the Cosmic Egg.* London: Lyrebird, 1973.

Pool, Christopher. *Olmec Archaeology and Early Mesoamerica.* Cambridge: Cambridge University Press, 2007.

Powell, Christopher. "The Shapes of Sacred Space: A Proposed System of Geometry Used to Lay Out and Design Maya Art and Architecture and Some Implications Concerning Maya Cosmology." Ph.D. dissertation, University of Texas, Austin, 2010.

Rogers, John H. "Origins of the Ancient Constellations: I, The Mesopotamian Traditions." *Journal of the British Astronomical Association* 108, no. 1 (1998).

———. "Origins of the Ancient Constellations: II, The Mediterranean Traditions." *Journal of the British Astronomical Association* 108, no. 2 (1998).

Sachs, Curt. *The Rise of Music in the Ancient World, East and West.* New York: Norton & Co., 1943.

Santillana, Giorgio de, and Hertha von Dechend. *Hamlet's Mill: An Essay Investigating the Origins of Human Knowledge and Its Transmission through Myth.* Boston, MA: David Godine, 1969.

Soustelle, Jacques. *The Olmecs: The Oldest Civilization in Mexico.* Norman: University of Oklahoma Press, 1985.

Segal, Michael. "The Jewish Calendar in Jubilees: A 364-Day Solar Year." Accessed June 9, 2017. http://thetorah.com/jewish-calendar-in-jubilees-a-solar-year (accessed November 28, 2017).

Sugiyama, Saburo. "Teotihuacan City Layout as a Cosmogram." In *The Archaeology of Measurement,* edited by Iain Morley and Colin Renfrew, 130–49. Cambridge: Cambridge University Press, 2010.

Terpstra, Siemen. "An Introduction to the Monochord." In *Alexandria: The Journal of the Western Cosmological Traditions.* Vol. 2, edited by David Fideler, 137–66. York Beach, ME: Red Wheel/Weiser, 1993.

Ulansey, David. *The Origins of the Mithraic Mysteries.* Oxford: Oxford University Press, 1989.

Van Buitenen, J. A. B, trans. *The Mahābhārata: 1, The Book of the Beginning.* Chicago, IL: University of Chicago Press, 1973.

West, M. L. *Hesiod: Theogony and Works and Days.* Oxford: Oxford University Press, 1988.

———. *The Orphic Poems.* Oxford: Oxford University Press, 1983.

Wolkstein, Diane, and Samuel Noah Kramer. *Inanna: Queen of Heaven and Earth.* New York: Harper & Row, 1983.

Wu Cheng'en. *Journey to the West.* Translated by W. J. F. Jenner. Beijing: Foreign Language Press, 1982.

Index